Death-Bed Visions

and other writings

Sir William Barrett (1844–1925) was a pioneering figure at the crossroads of science and the unseen. Born in Kingston, Jamaica in 1844 and died in London in 1925, Barrett was a distinguished physicist who co-founded the Society for Psychical Research (SPR) in 1882, the first organization devoted to the scientific study of paranormal phenomena. Initially skeptical of the supernatural, he underwent a dramatic shift in the 1870s after witnessing events he could not explain—such as a mesmerist's subject apparently divining his thoughts and mysterious "rapping" sounds that answered questions with uncanny intelligence. These encounters spurred Barrett to pursue psychical research with scientific rigor, leading him to convene like-minded colleagues and launch the SPR.

In parallel, Barrett's mainstream scientific career flourished. Appointed Professor of Experimental Physics at the Royal College of Science in Dublin in 1873, he pioneered work in acoustics and metallurgy—discovering the silicon-iron alloy "stalloy" used in early electrical engineering. His contributions earned him election as a Fellow of the Royal Society in 1899 and a knighthood in 1912. Barrett's life's work embodied a quest to unite empirical science with the exploration of consciousness and the unseen, reflected in the five works collected in this edition.

First Principal Edition
published by the Principle Press, Los Angeles
http://principle.press

Printed in the United States of America

ISBN (print) 9781967243129
ISBN (ebook) 9781967243136
Library of Congress Control Number 2025910317

First Printing, 2025

Death-Bed Visions
and other writings
Sir William Barrett, F.R.S.

published by the Principle Press
http://principle.press

PORTRAIT OF THE AUTHOR

CONTENTS

OTHER WRITINGS

Additional Material

ABOUT THIS COLLECTION

This new edition of *Death-Bed Visions and Other Writings* brings back into print Sir William Barrett's *Death-Bed Visions* (1926) along with four of his related shorter works – *Creative Thought and the Problem of Evil* (1914), *Swedenborg: The Savant & The Seer* (1912), his Presidential Address to the Society for Psychical Research (1904) and, serving as an epilogue of sorts, *Some Reminiscences Of Fifty Years' Psychical Research* (1924) – uniting these five writings in one volume for the first time. Though varied in form and focus, each of these works in its own way sheds light on Barrett's lifelong exploration of the unseen dimensions of experience and the interplay between scientific inquiry and spiritual insight; it is this underlying unity of purpose that prompted the Principle Press to unite them in a single collection. Editing this collection entailed several editorial decisions aimed at preserving the integrity of Barrett's work while enhancing its usability. Specifically, all of the author's original footnotes have been restored; the original editor's preface from the 1926 edition of Death-Bed Visions is reprinted; any internal cross-references have been updated to reflect this edition's pagination; the index from the first edition has been carefully reconstructed and expanded. The essays themselves have been carefully reconstructed from the first editions and combined here to one volume. Many of these additions, to our knowledge, have never been available in later editions. Through these efforts, we hope to honor Barrett's voice while making his insights accessible to a new generation of readers, thus setting the stage for a deeper engagement with his ideas in the pages that follow. In this new edition, *Death-Bed Visions,* and to some degree, *Creative Thought and the Problem of Evil* serve as centerpieces, illustrating Barrett's enduring effort to find empirical support for spiritual truths.

Sir William Barrett was a brilliant and unique scientist who truly stood at the threshold of two worlds. In the early 1920s, in the twilight of a long and distinguished career, Sir William Fletcher Barrett confidently affirmed his most controversial conviction: that human personality endures beyond death. "I am personally convinced that the evidence we have published decidedly demonstrates (1) the existence of a spiritual world, (2) survival after death, and (3) occasional communication from those who have passed over," Barrett stated, adding that it is "hardly possible to convey to others... an adequate idea of the strength and cumulative force of the evidence that has compelled [my] belief." Such words, spoken by an English physicist and Fellow of the Royal Society, capture the unique position Barrett occupied. Here was a *scientist* – professor of experimental physics in Dublin for nearly four decades – who was also a pioneering *psychical researcher* intent on probing the unseen dimensions of human existence. As a founder of the Society for Psychical Research (SPR) in 1882, Barrett straddled the boundary between the Victorian scientific establishment and the era's burgeoning interest in paranormal and spiritual phenomena. This new edition, *Death-Bed Visions and Other Writings*, brings together Barrett's 1926 classic *Death-Bed Visions* along with four other short works that will shed light on the philosophical, scientific, and spiritual concerns driving Barrett's life and thought. In these writings, we see a mind striving to reconcile empirical observation with spiritual insight, grounded in both scientific rigor and a profound spiritual curiosity.

Barrett's career offers rich biographical context for understanding the essays that follow. Born in 1844 and educated in the sciences, he made his mark in physics by discovering new alloys and phenomena (such as the magnetization-induced contraction of nickel). Elected an F.R.S. in 1899, Barrett was, by all accounts, a respected mainstream scientist. Yet from the 1870s onward he became equally absorbed by psychical research – the scientific investigation of experiences like telepathy, clairvoyance, and apparitions. He conducted early telepathy experiments (one 1876 paper on "phenomena associated with abnormal conditions of mind" was initially met with skepticism by the British Association) and became the *"prime-mover"* in establishing the SPR. In the SPR's first year he served as editor of its journal, and he would later serve as its President in 1904. Barrett approached psychic phenomena with open-minded empiricism: he was willing to hypothesize natural explanations (early on he attributed reports of levitation or spirit raps to hallucination), but he changed his views when faced with credible evidence under controlled conditions. Over decades of research, investigating everything from dowsing to mediumship, Barrett became ever more persuaded that the materialist view of mind was incomplete. His intellectual journey was not without struggle –

he "groped [his] way" through profound questions of human personality and its survival – but it was driven by a coherent purpose. As the works in this collection demonstrate, Barrett sought *a unifying truth*: that mind and spirit are integral parts of nature, accessible to study and bearing on humanity's deepest existential questions.

Perhaps the most enduring of Barrett's works, *Death-Bed Visions* (first published 1926, just after his death) is a pioneering study of extraordinary experiences reported at life's end. In this concise monograph, Barrett gathered and analyzed dozens of cases in which dying individuals – often in their final minutes or hours – appeared to see visions of deceased loved ones or otherworldly scenes. Such deathbed apparitions had long been noted anecdotally by doctors, nurses, and families, but Barrett's book constitutes *"the first known study of its kind,"* treating the phenomenon with careful documentation and inquiry. He approaches the subject as a scientist and sympathetic listener: noting that hallucinations are not uncommon in the dying, Barrett nonetheless highlights a subset of instances that defy simple explanation. In particular, he was fascinated by cases where the dying person's vision included someone believed to be alive – only for others to discover that person had in fact died previously, unknown to the dying individual. "These cases form, perhaps, one of the most cogent arguments for survival after death," Barrett wrote, "as the evidential value and veridical (truth telling) character of these visions of the dying is greatly enhanced when the fact is undeniably established that the dying person was wholly ignorant of the decease of the person he or she so vividly sees." In other words, if a person on their deathbed joyously calls out to a dear friend or relative who *nobody present* knew had already died, what could that imply? To Barrett, such occurrences strongly suggested that the "spirit form" of the deceased was truly present – a literal *meeting at the threshold* of this life and the next.

In *Death-Bed Visions*, Barrett's tone is compassionate, observant, and rigorously logical. He shares poignant stories of dying men, women, and even children who seemed to perceive more than this world alone. Yet he consistently asks the skeptical questions a scientist must. Could these be mere fevered imaginings, wish-fulfillments of a brain under duress? Barrett responds by sorting the cases into careful categories. Chapter I (Introduction) lays out the problem and the need for discrimination between genuine visions and hallucinations. He cites the eminent physiologist Charles Richet – a friendly skeptic who attributed many psychic phenomena to a mysterious subconscious sense ("cryptesthesia") rather than spirits – acknowledging that even Richet found deathbed cases *"disquieting"* and more explicable by a *"spiritist theory"* than by any mind-to-mind telepathy or chance coincidence. Barrett then proceeds through individual cases, giving pride of place to the most evidential instances: those where the apparition was not known to be dead by the dying person (and thus could not have been a preexisting expectation). Other chapters examine more typical visions of deceased persons known to have died, as well as related phenomena reported around deathbeds, such as mutual visions (e.g. a distant loved one seeing the dying person at the moment of death) and even musical sounds heard at a time of death. Throughout, Barrett writes as both a reporter and an analyst. He does not ask readers to accept every story uncritically, and instead invites the reader to judge whether any normal explanation suffices. Yet Barrett does not hesitate to make clear his own conclusion: taken as a whole, the consistency and sincerity of these visions, especially the surprises they sometimes contained, point to the reality of an afterlife. Even hardened skeptics, he notes, have at the very end of life shown "evidence that at the very end they knew there was an after-life."

It is worth remembering that *Death-Bed Visions* was published posthumously and remains incomplete by design. Barrett died in 1925 while still refining the manuscript. The published text includes an Editor's Preface (by his wife, Florence Barrett) explaining that Barrett had only sketched an introduction and planned to expand the discussion with references to other researchers like Ernesto Bozzano. Rather than have another hand finish his work, the editor chose to present Barrett's draft as-is. As a result, the book offers a tantalizing glimpse of Barrett's unfulfilled intentions. We see his keen interest in scientific critiques – he wanted to address why dying people rarely, if ever, hallucinate *living* persons if these visions are mere wishful thinking, and to explore analogies to well-documented cases of "traveling clairvoyance" or out-of-body experiences at death. He also gave thought to the psychology of children's visions that include imagery beyond their religious training, considering such cases unlikely to be "a mere flight of fancy." Though *Death-Bed Visions* the book is slim, *Death-Bed Visions* the subject was, for Barrett, vast and profound – "that little-known country" to which, as the editor movingly writes, Barrett himself "passed suddenly" in mid-stride, even as his thoughts so often *"took wing"* toward it. This new edition invites readers to share in that journey with Barrett as he stands at the border of life and death, attempting to map what lies beyond.

Barrett's little book would become *"something of a classic in the field,"* and its influence resonates today in modern near-death studies and hospice care literature. In Barrett's era, such accounts of deathbed visions were often treated with polite skepticism or quietly brushed aside as hallucinations. Yet Barrett saw significance where others saw taboo, carefully documenting cases of the dying seeming to greet unseen loved ones. Over a century later, mainstream science and medicine are finally catching up. The once-marginal topic of deathbed visions has gained widespread recognition, even in major media: *The New York Times Magazine* recently devoted prominent features to these phenomena, including Phoebe Zerwick's "What Deathbed Visions Teach Us About Living" (March 2024) and a follow-up piece, "Readers Share Stories of Their Loved Ones' Deathbed Visions" (April 2024). Zerwick's article highlighted research by hospice physician Dr. Christopher Kerr, who found through systematic interviews that an astonishing 88% of his terminal patients reported vivid end-of-life dreams or visions, usually involving deceased relatives and bringing profound comfort. The public response was telling: the Times was inundated with readers' own testimonies of bedside visions – deeply personal stories that, taken together, affirm how common and meaningful these experiences can be. Such open discussion in a mainstream forum would have been unthinkable in Sir William's day. What was once whispered about in Victorian sickrooms or dismissed by physicians as mere delirium is now the subject of serious conversation about the dying process. Indeed, hospice caregivers today widely recognize these vision experiences as a natural and even beneficial aspect of end-of-life care, a marked shift toward empathy and curiosity that surely would have gratified Barrett. As one modern commentator noted, when Dr. Kerr's findings went public, nurses around the world were *"grateful that someone in the medical field validated what they have long understood"* – namely, that deathbed visions are real, often transcendent encounters that help the dying find peace. In our current culture, exploring these final mysteries has moved from the fringes to the mainstream, reflecting a broader openness to understanding what the end of life may hold in terms of human experience and consciousness.

This growing acceptance has been paralleled by a surge in scientific research into near-death experiences (NDEs) and related phenomena, bringing empirical rigor to questions Barrett could only approach through anecdote. In the decades since psychiatrist Raymond Moody first coined the term "near-death experience" in 1975, physicians and academics have amassed a substantial body of data suggesting that consciousness can manifest in extraordinary ways at the brink of death. Pioneering researchers such as Dr. Bruce Greyson have catalogued common elements of NDEs – out-of-body perception, journeys through a tunnel or light, panoramic life review – and shown their life-changing aftereffects on those who experience them. Prospective hospital studies have lent credibility to what was once the stuff of mysticism. For example, Dutch cardiologist Pim van Lommel's famous 2001 study in *The Lancet* reported that 18% of cardiac arrest survivors described an NDE, and intriguingly, he found no clear physiological cause (such as oxygen deprivation or drug effects) to predict which patients would have these experiences. More recently, critical-care physician Sam Parnia led the AWARE II study (2017–2020) across multiple hospitals to objectively monitor the brain and consciousness during cardiac arrest. The results, published in 2022, were striking: about one in five patients who were resuscitated from prolonged cardiac arrest later reported lucid experiences of death – such as a feeling of leaving the body or a meaningful evaluation of one's life – even though they had been clinically unresponsive at the time. What's more, Parnia's team detected surges of brain waves (including gamma frequencies associated with awareness) up to an hour into CPR, long after the point when the brain is typically assumed to have shut down. These findings hint that the "window" of consciousness may not slam shut as neatly at the moment of death as once supposed. "Our results offer evidence that while on the brink of death and in a coma, people undergo a unique inner conscious experience," Parnia observes, noting that such lucid episodes cannot be written off as "a trick of a disordered or dying brain." Even phenomena like terminal lucidity – the uncanny return of clarity and memory in some dementia patients just before death – are now drawing scholarly attention, precisely because they defy our neurological expectations of irreversible decline

Researchers acknowledge that these end-of-life moments of lucidity *"contradict what we know about the human brain,"* representing a genuine paradox in need of explanation. Across the board, today's investigations are driven by the same impulse that animated Barrett's work: a desire to examine extraordinary end-of-life experiences with scientific seriousness, open-minded curiosity, and humane respect for what they might reveal. Barrett, a physicist by training and a psychical researcher by calling, exemplified this balanced approach in an age when it was far easier to dismiss such reports. He was motivated by the conviction that truth is worth seeking even in areas that challenge orthodox theory. That ethos lives on. Modern research into NDEs and deathbed visions, often published in reputable medical and psychology journals, continues very much in Barrett's spirit – bridging the clinical and the existential, and probing whether consciousness might extend beyond the boundaries we once assumed. In this sense, *Death-Bed Visions* was not only a collection of intriguing cases; it was the beginning of a conversation that science is finally mature enough to carry forward.

The second work included here, *Creative Thought and the Problem of Evil* (1914), reveals Barrett in a more speculative and philosophical mood, wrestling with questions about the nature of mind, matter, and meaning that feel strikingly ahead of his time. In this extended essay, published a few years before *Death-Bed Visions*, Barrett ventured beyond documenting phenomena and into the realm of ideas – grappling with big-picture issues that would not look out of place in a 21st-century symposium on consciousness. A century before today's debates over whether machines might think or whether neuroscience can fully explain awareness, Barrett was already asking: *What is the relationship between human thought and the wider cosmos? Is mind a mere epiphenomenon of matter, or could it be fundamental to the fabric of reality?* His exploration synthesizes insights from the science of his day (such as early theories of the subconscious and even telepathy) with a deeply reflective strain of spiritual philosophy. Notably, Barrett argues for a vision of the universe suffused with mind and purpose. Anticipating ideas that modern philosophers and physicists are revisiting, he suggested that the material world we perceive is the *product* or "offspring" of an unseen creative intelligence. "The Ultimate Reality must be Mind," Barrett insists, positing an infinite "Intelligence behind the universe" intimately linked with our own conscious selves. In an era dominated by mechanistic science, this was a bold assertion – but today, as scholars debate *panpsychism* (the idea that consciousness is a fundamental feature of matter) or propose that physical reality may emerge from informational or mental substrates (or that our own reality could even be a computer simulation), Barrett's once-radical idea sounds uncannily prescient. He even confronted the age-old "problem of evil" within this framework of a mindful universe. If a *Divine Mind* underlies nature, why do we find so much suffering, chaos, or cruelty in the world? Barrett acknowledged the force of this question, noting how "much in Nature appears to us imperfect, objectionable, and evil" when viewed in isolation. His response drew on the notion of our limited vantage point: just as a microbe or a momentary pain can play its part in a larger biological good, what we call "evil" might reflect our fragmentary grasp of a vastly larger, ultimately coherent cosmic tapestry.

One need not share Barrett's Edwardian spiritual optimism to see how forward-reaching his concerns were. Today, discussions about the "hard problem" of consciousness – the enigma of how subjective experience arises from the brain (if indeed it does) – and renewed cross-disciplinary dialogues on science and spirituality echo many of Barrett's themes. The rise of artificial intelligence has further intensified these questions: as we design machines that mimic human cognitive functions, we are forced to examine what, if anything, distinguishes algorithmic processing from the creative insight of a conscious mind. Neuroscience, for all its breathtaking progress in mapping brain correlates of thought, still struggles with explanatory gaps when it comes to the first-person reality of mind. And philosophers continue to debate whether our universe is best understood as fundamentally mental, physical, or something in between. In *Creative Thought*, Barrett was grappling with precisely this nexus of science, consciousness, and metaphysics long before it was fashionable. He brought to it the same rigor and wonder that mark his work on deathbed visions. In doing so, Barrett set out ideas that would remain on the intellectual agenda well into the next century: the possibility that mind is more than an emergent trick of biology, the search for meaning in a seemingly indifferent cosmos, and the hope that expanding our understanding of consciousness might illuminate even the darkest problems – evil, suffering, and death – that humanity faces. The dialogue he began continues today, making his reflections in 1914 as thought-provoking now as they must have been then.

Creative Thought and the Problem of Evil (1914), reveals Barrett in a more speculative and philosophical mood. If *Death-Bed Visions* grapples with evidence for *life after death*, *Creative Thought* grapples with the nature of *life and mind during life* – and how a world governed by a divine Mind can contain suffering and evil. Written on the eve of World War I (a context that undoubtedly sharpened the problem of evil for many thinkers of the time), Barrett's essay ranges over science, psychology, and theology in search of answers. He was not the first to tackle the age-old problem of evil (the question of how pain and moral evil can exist under an all-good, omnipotent Providence), but Barrett brings a distinctive perspective rooted in his scientific understanding of evolution and mind. In *Creative Thought and the Problem of Evil,* Barrett posits that mind – a creative, directive intelligence – is an immanent force in the universe, guiding evolutionary progress yet allowing for freedom and growth. "The psychic factor in evolution is more than immanent Power; it is immanent Thought which is the mode of the Divine indwelling in nature," Barrett asserts. Here we see Barrett's spiritual philosophy in a nutshell: nature is not a blind clockwork, but suffused with *creative thought* emanating from the Divine. This idea echoes currents in early 20th-century thought (for example, the "Creative Evolution" of Henri Bergson, whom Barrett references) and even anticipates Teilhard de Chardin's later vision of a universe unfolding towards mind. Barrett, as both physicist and psychical researcher, was convinced that mind is a fundamental reality – one that conventional science had underestimated.

Barrett's emphasis on "creative thought" leads him to a nuanced take on evil. If a divine Mind indwells nature, why would that Mind permit wrongdoing or misery? Barrett suggests that the creative advance of the world necessitates genuine freedom and struggle. In his view, the Divine Intelligence works organically rather than coercively: it endows creation with a degree of autonomy, a capacity to err and learn, which is essential for the development of virtue, knowledge, and individuality. This framework aligns with the classic "free will defense" in theodicy (theology of evil) – the idea that creatures must be free to choose, even choose wrongly, for goodness to have meaning – but Barrett bolsters it with insights from psychology and paranormal phenomena. The text delves into examples of the mind's latent powers (the "psychic factor") as evidence of our spiritual endowment. For instance, Barrett discusses hypnosis, faith-healing, and even stigmatic phenomena (the mysterious appearance of wounds corresponding to Christ's crucifixion) as indications that thought can affect physical reality. Such phenomena, he argues, hint at a higher aspect of the human mind that connects with the Divine Mind. If properly understood, creative thought could be a means to *overcome evil* – by harnessing spiritual resources to heal, to inspire moral progress, and to find meaning in suffering.

The style of *Creative Thought and the Problem of Evil* reflects Barrett's role as both scientist and seeker. He writes in a scholarly yet accessible tone, citing contemporary thinkers (the common terms in the text range from "Bergson" and "Leibnitz" to "unconscious" and "subliminal") and using analogies from physics and biology. Yet he also speaks from the heart of a spiritually engaged person wrestling with ultimate questions. At one point, Barrett invokes the mystery of consciousness – how matter has given rise to mind – as a clue that creation is driven from within by an intelligent principle. The universe, he proposes, is an expression of "the Divine indwelling", and our own creative efforts (in art, science, love, etc.) partake of that wider Creativity. Evil, in this picture, is not an irredeemable blight but a byproduct of a world in process – a world where finite minds gradually awaken to their unity with the Divine will. Barrett does not claim to have solved the riddle of evil once and for all. Rather, he offers a hopeful *framework*: if mind is truly *first* in the universe, then purpose and goodness are its deepest laws, and evil is ultimately a problem to be resolved by the expansion of conscious creative thought. As readers will see, this essay provides invaluable insight into Barrett's *personal faith*. Unlike some psychical researchers of his era who embraced a sterile Spiritualist creed, Barrett's spirituality was intellectually rich and deeply ethical, concerned with aligning the discoveries of psychical science with a moral philosophy of life. *Creative Thought and the Problem of Evil* thus illuminates the optimistic idealism underlying Barrett's empirical work – the belief that expanding human knowledge of the unseen will contribute to *overcoming ignorance and evil.* It complements *Death-Bed Visions* by addressing a logical question: if there is indeed a benign spiritual world and survival after death, how shall we understand the struggles of this world? Barrett's answer is an invitation to see our lives as part of a larger, divine creative adventure, not a meaningless chaos.

A further key to Barrett's mind is found in *Swedenborg: The Savant & The Seer* (1912), his appreciative study of Emanuel Swedenborg. Swedenborg (1688–1772), the Swedish scientist-turned-mystic, had become legendary as a man who bridged *two realms*: he was renowned for his contributions to mining, astronomy, anatomy, and physics before midlife, and for his extensive visionary journeys in the spiritual world thereafter. It is not hard to see why Barrett would be drawn to write about him. In Swedenborg, Barrett recognized a kindred spirit – *"a pioneer and leader both in philosophy and theology,"* as one speaker at the 1910 Swedenborg Congress put it. Barrett himself contributed to that international congress marking the centenary of the Swedenborg Society, and the substance of his remarks became the basis for this essay. His aim was to introduce Swedenborg to a new generation, highlighting the uncanny way Swedenborg's scientific insights and spiritual revelations converged.

Barrett's portrait of Swedenborg is admiring yet measured. He recounts Swedenborg's early achievements as a "master of all the older sciences" who even "led the way into other sciences quite new to the student." Indeed, Barrett notes that Swedenborg anticipated modern ideas – from nebulous star formation to the concept of neurons – with a clarity that would only be appreciated long after his time. But the crux of Barrett's essay lies in Swedenborg's transformation into a seer. Beginning in 1744, Swedenborg claimed to have his spiritual eyes opened, enabling him to communicate with spirits and angels and to explore heaven and hell as consciously as one might explore a foreign country. To a 20th-century physicist like Barrett, such claims could easily invite skepticism. Yet Barrett approaches them with both respect and curiosity. He emphasizes the remarkable consistency and lucidity of Swedenborg's reports, as well as the noble theology they revealed – a vision of a universe defined by *Divine Love and Wisdom*, where the material world corresponds to spiritual realities. Barrett reminds readers that many eminent figures in literature and philosophy (from Coleridge and Elizabeth Barrett Browning to Emerson and William James Sr.) drew inspiration from Swedenborg's cosmic Christianity. As Barrett boldly states, *"it is hardly possible to exaggerate Swedenborg's mental power, or to place him too high among the selectest company of men."* Coming from Barrett, himself a man acquainted with some of the greatest minds of his age, this is indeed high praise.

What does Barrett extract from Swedenborg's example? Beyond the biographical facts, Barrett uses Swedenborg to underscore the harmony of scientific truth and spiritual truth. He points out that Swedenborg never repudiated reason or observation; even in his visionary theology, Swedenborg insisted that his spiritual experiences were *real events* open to rational scrutiny, not mere dreams. In an era when many scientists saw a gulf between the measurable world and matters of faith, Swedenborg (and by extension, Barrett) offers a counter-model: the scientist-mystic who finds continuity between exploring nature and exploring the spirit. Barrett's essay draws subtle parallels to psychical research – noting, for instance, Swedenborg's occasional veridical psychic perceptions (such as his famous vision of a Stockholm fire while in Gothenburg, later confirmed as accurate) as early evidence of the phenomena the SPR was systematically studying. While careful not to present Swedenborg uncritically (Barrett acknowledges that not every claim Swedenborg made can be substantiated), he clearly finds in Swedenborg an encouraging precedent for his own endeavors. If one of the towering intellects of the Enlightenment could devote his later years to charting the afterlife, might not Barrett's contemporaries take seriously the SPR's more modest attempts at doing the same? In tone and content, *Swedenborg: The Savant & The Seer* serves as Barrett's bridge to the past – a way of placing his cutting-edge Victorian/Edwardian research into a longer historical continuum of inquiry into the unseen. It reminds us that the questions Barrett pursued were not *novelties* of spiritualism or fin-de-siècle occultism, but part of a lineage of human inquiry that included great minds of earlier ages. This contextual, historical perspective enriches the reader's understanding of Barrett's own motivations. Just as Swedenborg sought to unify his era's knowledge with transcendent revelation, Barrett sought to fuse modern science with eternal spiritual truths. The inclusion of this piece in the collection underscores how Barrett's outlook was informed by those who came before – it places his life and thought in a broad, inclusive frame that spans from the 18th century to what was, in 1912, the very forefront of psychical science.

The final two pieces in this volume – *Sir William Barrett's Presidential Address to the SPR*, delivered in 1904, and *Some Reminiscences of Fifty Years' Psychical Research* (1924) – offer a fitting capstone, revealing a seasoned researcher reflecting on decades of a life's work. In these retrospections, we encounter Barrett in an explicitly editorial and personal voice, distilling the lessons of a lifetime. His 1924 address in particular does not merely recount personal memories – it uses them to illustrate the evolution of a field and to argue for its significance. Barrett speaks as a scientist who has seen strange things and who urges the world to take those things seriously.

A major theme of the two addresses is the complexity of human personality. Decades of experiments and investigations had convinced Barrett that the mind cannot be fully explained by brain physiology alone. He references cases of telepathy, clairvoyance, apparitions, and haunted houses, weaving them into a narrative of discovery. Many of these phenomena, he notes wryly, were once dismissed as superstition yet have "forced themselves upon our attention" through careful research. Barrett uses his platform as SPR President to call for open-minded yet disciplined study – a "*bridge between the physical and psychical*," as he put it, must be built with the tools of science but the vision of philosophy. He cites examples of rigorous investigations: for instance, the SPR's famous *Census of Hallucinations* that gathered thousands of accounts of death coincidences and apparitions, or the controlled telepathy trials he and colleagues conducted with marked success. These, Barrett argues, suggest that human consciousness has capabilities (and perhaps extensions) that transcend our ordinary senses. By sharing his own reminiscences – including early struggles to be heard by scientific bodies and later triumphs when skeptics were won over by evidence – Barrett personalizes the history of psychical research. Readers gain an appreciation for the bold, and sometimes lonely, path he chose in the 19th century, when pursuing such topics could imperil one's reputation.

Most powerfully, Barrett's presidential retrospectives reiterate the conclusion we saw at the outset: that the weight of evidence *compels belief* in a spiritual dimension. In his 1924 recollection, he unabashedly concludes that there is *"convincing evidence"* for an afterlife and communication from the departed. The Presidential Address in 1904 lays the groundwork for that conclusion, as Barrett outlines phenomenon after phenomenon that point to the same hypothesis.

He addresses counterarguments – for example, could telepathy between living people explain some cases instead of spirits? Yes, possibly in some instances, Barrett admits, but then what of the cases telepathy *can't* cover, such as the deathbed visions where no living mind knew the information? In those, he sees proof positive of discarnate consciousness. He even entertains the philosophical implications: if an intelligent order of beings exists unseen alongside us, what does that say about the universe? His scientific mind concedes that we are only "groping our way" toward a full theory, but his human heart is warmed by the notion that psychical research has provided *"glimpses of the beyond"* sufficient to bolster faith in immortality. At the same time, Barrett cautions against credulity. He recounts how he once debunked fraudulent mediums and how the SPR's investigations have exposed cases of trickery. Such experiences taught him to demand high standards of evidence – a legacy the SPR carried forward. Thus, Barrett's retrospective address strikes a balance between extraordinary claims and empirical scrutiny. It is both a celebration of how far the research had come and a sober reminder that much remained unknown.

One is also struck by Barrett's combination of spiritual conviction and scientific humility in delivering these addresses. He was, by 1904, a man of faith (a Christian with a spiritualist lean) and of science, and he saw no conflict between the two when each is properly understood. In one stirring passage, Barrett invokes the metaphor of the "unseen world" as something like the luminiferous ether of physics – an invisible medium pervading space. Just as physicists of his day believed the ether could host light waves, Barrett believed the spiritual world might interpenetrate our own, carrying thoughts and souls. *"It is in harmony with all we know,"* he wrote in a later work, *"to entertain a belief in an unseen world, in which myriads of living creatures exist... and it is possible that the evolutionary development of such a world has run on parallel lines to our own."* His addresses often return to this vision: that we are citizens of a two-tiered reality, physical and spiritual, and that future science will expand to fully include the latter. When Barrett stepped down from the SPR presidency, he left a legacy of pushing against the boundaries of official science, expanding its scope to include the human spirit. The inclusion of his Presidential Address in this collection allows modern readers to hear Barrett's voice of experience – a voice that is candid, erudite, occasionally wry, but ultimately inspiring in its steadfast belief that *truth is larger than the visible world.*

Sir William Barrett's hope that science would one day validate an "unseen world" coexisting with the physical finds both echoes and challenges in today's research. Mainstream science remains largely grounded in materialism, yet a growing number of scientists and philosophers are seriously examining whether consciousness and subjective experience might not be fully reducible to brain activity. This inquiry spans neuroscience, quantum physics, and philosophy of mind, reflecting an ongoing effort to bridge Barrett's two-tiered reality – physical and spiritual – in a rigorous, evidence-based way.

Research into consciousness has exposed the limits of a strictly materialist worldview. Neuroscience, for instance, has made great strides mapping brain activity, but the mind-brain enigma persists. The infamous "hard problem" of consciousness – how subjective awareness arises from matter – remains unsolved. In fact, many experts admit that purely examining neural circuits may never fully explain first-person experience. Recent analyses have urged a more critical look at the assumption that mind is *only* an emergent brain product. When considered together, various empirical findings "give us good reasons to look upon the physicalist assumptions with a more critical eye," suggesting that models treating consciousness as a fundamental aspect of reality deserve attention. For example, some neuroscientists have noted cases of vivid mental experience under minimal brain function (such as near-death experiences during cardiac arrest), which hint that consciousness might at times operate independently of normal brain activity. While controversial, such data have prompted calls for "post-materialist" science that can accommodate phenomena of mind beyond neural mechanisms. At the very least, there is growing acknowledgment that our current neurobiological models do not fully capture the depth of subjective life.

Quantum physics has also provided unexpected avenues to revisit Barrett's vision. Although most physicists do not invoke "spirit" explicitly, some interpretations of quantum theory raise profound questions about the role of observers and consciousness in reality. Notably, a minority of researchers propose that quantum processes might be essential to consciousness. A 2024 study on how anesthetics induce unconsciousness found that anesthetic drugs act on quantum-level structures (microtubules) in brain neurons – evidence which "supports the quantum model of consciousness" according to the lead neuroscientist. Such findings revive an idea that mind may be tethered to the fabric of the universe in ways classical biology alone cannot explain. As one researcher put it, embracing a quantum perspective on mind could usher in "a new era in our understanding of what we are," potentially linking us to the cosmos in a more "holistic" way. While far from proving a spiritual medium, quantum-consciousness research broadens the scientific imagination toward an interconnected reality – a step in the direction Barrett imagined.

In philosophy of mind, the debate over materialism versus alternative paradigms has intensified. A significant minority of philosophers now favor non-physicalist or dual-aspect frameworks for consciousness. Ideas like panpsychism – the view that consciousness is a fundamental feature of matter – have moved from fringe to serious discussion in academic circles. Prominent neuroscientists and philosophers (e.g. Christof Koch and David Chalmers) have explored whether consciousness might pervade the universe in some form, an outlook reminiscent of Barrett's two-tier reality. Importantly, multiple competing theories of consciousness are being tested today – from Global Neuronal Workspace Theory (a purely neural-circuit model) to Integrated Information Theory (which intriguingly implies that even simple systems possess some degree of consciousness). A recent large-scale experiment pitting such theories against each other ended in a draw, underscoring that no single materialist theory yet accounts for the emergence of mind. Some scholars even argue that certain complex aspects of experience might point to a deeper, as-yet unidentified principle in nature. In parallel, the rise of advanced artificial intelligence has added new dimensions to these questions. If human-like AI were to exhibit signs of understanding or awareness, would that confirm consciousness as a purely computational phenomenon? Or would it suggest that genuine subjective experience requires something more (some immaterial spark)? Researchers remain divided. Many cognitive scientists hold that if the brain is an organic machine, then sufficiently sophisticated machines could also be conscious. Others contend that qualia (the raw feel of experience) entail more than information processing, aligning with the view that mind taps into a fundamental layer of reality that AI might lack. This debate is sharpening our philosophical and scientific definitions of consciousness, forcing us to ask whether awareness is ultimately an emergent neural process or a glimpse of an "unseen" ontological realm.

How far have we come toward Barrett's vision of a spiritually inclusive science? Measured against the early 20th-century optimism that the "ether" would soon reveal invisible worlds, progress is cautious but notable. On one hand, empirical science has not confirmed any literal otherworldly substance or definitively proven that consciousness can exist independent of the brain. The prevailing scientific paradigm is still one of methodological naturalism – every phenomenon must be explained, or at least correlated, with physical processes. Skepticism remains high toward overtly "spiritual" hypotheses, partly due to the lack of reproducible evidence and the successes of neuroscience in linking mind to brain. On the other hand, the frontiers of knowledge have undeniably expanded in directions that Barrett would find encouraging. Consciousness is no longer dismissed as a trivial by-product; it is recognized as a profound mystery at the heart of our understanding of nature. Serious researchers convene at interdisciplinary conferences (such as *The Science of Consciousness*) to probe *"the nature of existence and our place in the universe"* through both "neurocomputational" models and "funda-mental" approaches that treat consciousness as intrinsic to the cosmos. Importantly, a body of reputable work – from peer-reviewed studies of mind-brain correlations to theoretical physics papers – is exploring the possibility that mind and matter are deeply intertwined aspects of reality. These developments represent tentative strides toward integrating subjective experience (and even notions of spirituality) with scientific inquiry.

Today, researchers studying phenomena like near-death experiences, psi abilities, or mind-matter interactions still often struggle with reproducibility and methodological criticism. The interpretations of these phenomena are contested – for example, whether a burst of gamma waves in a dying brain signifies the departure of a soul or just a last neural flurry. Establishing frameworks that allow "spiritual" hypotheses to be tested without bias is another frontier. Additionally, integrating any form of consciousness principle into physics poses theoretical hurdles: how would an immaterial mind influence physical processes without violating known laws? Despite these challenges, the conversation has opened up. There is a cautious openness in pockets of the scientific community to consider that reality might have dimensions not captured by classical materialism. The philosophy of mind's continued debate, coupled with cutting-edge experiments (in neuroscience labs and quantum physics alike), keeps Barrett's question alive. Modern science has not (yet) confirmed that we are "citizens" of a literal two-tiered reality in the way Sir William Barrett envisaged. However, it *has* broadened its scope to seriously entertain the hardest problems of consciousness and acknowledge phenomena that hint at deeper layers of reality. This represents a meaningful shift from a century ago: an emerging willingness to probe the "unseen" with scientific rigor, even if definitive answers remain elusive. The journey continues, grounded in fact and open in spirit, much as Barrett hoped.

Taken together, the four works in this volume present a multifaceted portrait of Sir William F. Barrett – not only as a man of his time, but as one whose ideas remain intriguing and relevant. From the detailed case studies of *Death-Bed Visions* to the expansive metaphysics of *Creative Thought,* and the historical analysis of the Swedenborg essay and the personal testimony of the SPR address, we see an intellect striving for *synthesis*. Barrett sought a synthesis of *science and spirituality*, of *reason and faith*, of *observation and intuition*. His life's work was driven by the conviction that reality, at its heart, is unitary – that the seen and unseen are part of one continuum governed by divine law. In Barrett's era, this was a bold stance: the growing dominance of materialistic science had led many to dismiss spiritual questions as obsolete. Barrett's response was not to reject science, but to extend it. He believed the tools of careful observation, experimentation, and logical inference could – and should – be applied to phenomena of the mind and soul. In this sense, he was a true successor to the Enlightenment's natural philosophers, *and* to visionaries like Swedenborg. Both the empiricist and the mystic lived in him.

For the modern reader, Barrett's writings offer both historical insight and timeless inspiration. Historically, they immerse us in the intellectual climate of the late 19th and early 20th centuries, when séances in dark parlors coexisted with the light of new scientific discoveries. We see Barrett painstakingly attempt to remove the stigma of the "occult" from psychical research and give it academic credibility – a struggle that continues, in some form, to this day. We also see how Barrett's Victorian faith infuses his work: his Christianity was broad and philosophically open, and he was comfortable citing ancient scriptures and contemporary poets alongside laboratory results. This breadth allowed him to engage with the "problem of evil" at a level that pure data could not reach, and to appreciate the spiritual genius of a figure like Swedenborg while still wearing a scientist's hat. In an age when specialized silos of knowledge often keep science and humanities apart, Barrett's integrative approach is refreshing. His work invites us to be unafraid of crossing disciplinary boundaries in pursuit of big truths.

Timeless inspiration shines through especially in Barrett's steadfast optimism. He was not naïve – he knew human folly and saw many promising leads turn to disappointment – but he maintained a fundamental hopefulness about what human beings could discover and become. He delighted in the "marvels and mysteries" he encountered, not because they were fantastical but because he felt they hinted at a *higher destiny for the soul*. In the closing of his 1924 reminiscences, Barrett reaffirms that the investigations of the SPR had, in his view, proven the reality of the spiritual world and the soul's survival. It is a conclusion he drew carefully, after decades of inquiry. We today might debate his interpretation of the evidence – indeed, many of Barrett's contemporaries did – yet there is no denying the courage and sincerity of his belief. It stands as a challenge to our cynical age: *what if* he was right? What if the "cumulative force of the evidence" is truly pointing somewhere profoundly important? Barrett invites us to consider that possibility with the utmost seriousness.

In preparing this modern edition, the Principle Press has aimed to let Sir William Barrett's work speak for itself, with only gentle guidance. The role of this introduction has been to provide context and connection – to frame each work so that the reader can appreciate how Barrett's interests and the topics at hand coalesce into a coherent worldview. As you delve into the writings that follow, you will notice recurring themes and language: Barrett's frequent references to *"unseen realities,"* his concern with *"proof"* and *"evidence,"* his use of terms like *"percipient," "phantasm,"* or *"cryptesthesia"* from the lexicon of early psychical research. You will also notice his humane perspective – whether describing a child's radiant smile at an angelic vision or pondering the moral development of souls, Barrett writes with a warmth and dignity that reflect his character. This collection is, in essence, the intellectual legacy of a man who devoted his life to exploring the invisible world in a methodical, enlightened, and empathetic way.

Nearly a century after Barrett's passing, the boundary between science and spirituality is still a frontier. Contemporary research into near-death experiences, consciousness, and anomalous cognition continues to raise many of the same questions Barrett once did. Readers may find that Sir William's essays resonate in surprising ways with current debates – be it the nature of consciousness, the integration of quantum theory and cosmology with concepts of a life principle, or the perennial human search for meaning in the face of mortality. Sir William Barrett's enduring contribution was to show that these big questions can be approached rationally without dispelling their wonder. In the end, Barrett, who died literally mid-research, might be seen as taking his own place among those he studied – one more soul embarked into that "undiscovered country." *Death-Bed Visions and Other Writings* allows us to accompany him partway, guided by his words and experiences. As we turn now to Barrett's own voice in the pages ahead, we do so with the understanding – imparted by this remarkable scientist-seer – that truth may lie on both sides of the veil, and that the pursuit of knowledge regarding our spiritual nature is a journey well worth taking.

Sources:

Barrett, William F. *Death-Bed Visions.* London: Methuen, 1926 (posthumous). [Public domain text; see especially Barrett's introduction and editor's preface for Barrett's aims and observations].

Barrett, William F. *Creative Thought and the Problem of Evil.* London: John M. Watkins, 1914. [Barrett's essay linking a divine creative principle with the existence of evil; key concept of "immanent Thought" as divine indwelling].

Barrett, William F. *Swedenborg: The Savant & the Seer. Transactions of the International Swedenborg Congress,* London, 1910. [Barrett's address praising Swedenborg's intellect and spiritual insight].

Barrett, William F. *SPR Proceedings 18* (1904): Barrett's "Presidential Address" [discusses skepticism and research difficulties]

Barrett, William F. *Address by the President: Some Reminiscences of Fifty Years' Psychical Research. Proceedings of the SPR* 18 (1904) and 34 (1924). [Barrett's reflections on decades of psychical research, asserting evidence for a spiritual world].

Encyclopedia of Occultism and Parapsychology. "Barrett, Sir William Fletcher (1844–1925)." Gale, 2001. (Provides biographical context and summaries of Barrett's contributions).

Proceedings of the Society for Psychical Research, Vol. XXXIV (1924–1925), containing *"Some Reminiscences of Fifty Years' Psychical Research"* on pp. 275–297. *Internet Archive* (State Library of Pennsylvania collection), 2015.

SPR Bibliography – *Proceedings* references for Barrett's works: *1904 Presidential Address (Proc. SPR Vol. 18)*; 1924 retrospective article (Proc. SPR Vol. 34).

Wikipedia. "William F. Barrett." (Accessed 2025). (Basic biographical data on Barrett's scientific career and SPR involvement.)

A FOREWORD TO DEATH-BED VISIONS: SIR WILLIAM BARRETT'S PIONEERING INVESTIGATION OF END-OF-LIFE EXPERIENCES

Sir William Barrett's posthumously published work *Death-Bed Visions* represents a landmark study in the field of psychical research, documenting phenomena that continue to intrigue scientists, healthcare professionals, and the general public alike. First published in 1926, the year after Barrett's death, this meticulous collection of accounts provides a window into the mysterious experiences reported by individuals approaching death. The book offers compelling narratives of dying persons who claim to see deceased relatives, spiritual entities, and otherworldly environments – often with remarkable clarity and conviction that transcends what many would dismiss as hallucination.

Sir William Fletcher Barrett (1844-1925) was born in Jamaica, where his father William Garland Barrett served as a Congregationalist minister with the London Missionary Society. The family returned to England in 1848, settling in Royston, Hertfordshire, before moving to Manchester in 1855, where Barrett attended Old Trafford Grammar School. Despite financial limitations that prevented formal higher education, Barrett's scientific aptitude led him to attend lectures on chemistry and physics, eventually becoming an assistant to physicist John Tyndall at the Royal Institution.

Barrett's career in physics was distinguished and productive. In 1873, he was appointed Professor of Experimental Physics at the Royal College of Science for Ireland, a position he held until his retirement in 1910. His scientific contributions were significant – he discovered Stalloy, a silicon-iron alloy essential to early telecommunications technology. In 1882, he documented the shortening of nickel when magnetized, another important observation in materials science. Barrett's scientific accomplishments earned him election to the Royal Society in 1899 and knighthood in 1912.

Late in his career, Barrett married Florence Elizabeth Perry in 1916, a pioneering woman in medicine who co-founded the London School of Medicine for Women. Barrett died in London on May 26, 1925, having left an indelible mark on both physics and psychical research.

While respected for his contributions to physics, Barrett's enduring legacy may be his pioneering work in psychical research. His interest in paranormal phenomena began in the mid-1860s when he visited Irish physician and mesmerist John Wilson. Barrett's scientific approach to investigating unexplained phenomena set him apart from both uncritical believers and dogmatic skeptics.

In 1882, Barrett played a crucial role in founding the Society for Psychical Research (SPR) after calling a meeting at the offices of the British National Association of Spiritualists. He later helped establish the American Society for Psychical Research in 1885. Throughout his career, Barrett investigated telepathy, clairvoyance, apparitions, mediumship, dowsing, and – most relevant to our current focus – deathbed visions.

Death-Bed Visions was published by Methuen & Co. Ltd. in London in 1926, the year after Barrett's death. This 116-page volume represents the culmination of nearly 50 years of psychical research. The book is historically significant as "the first known study of its kind" on what are now termed "deathbed phenomena."

Barrett's work focuses primarily on a specific type of deathbed experience: cases where dying individuals see and recognize deceased relatives or friends, sometimes without knowing that these individuals had died. Barrett considered this particular phenomenon especially compelling evidence for survival after death, as it could not be easily explained as wishful thinking or suggestion.

One of the most striking accounts in the book involves a woman identified as Mrs. B., who was in labor at the Mothers' Hospital in Clapton and suffering from serious heart failure. Lady Barrett, who was an obstetrical surgeon at the hospital, documented the case in detail. After delivering her baby, Mrs. B. began to experience visions, first commenting on increasing darkness before suddenly exclaiming with joy about seeing "lovely brightness" and "wonderful beings."

What makes this case particularly evidential is that Mrs. B. recognized her deceased father among these beings, saying, "Why, it's Father! Oh, he's so glad I'm coming; he is so glad." She maintained awareness of both her physical surroundings and the visionary realm until her death, even arranging practical matters regarding her premature baby's care while continuing to reference the "lovely brightness."

Another compelling case Barrett documented involves a dying young woman who, in the presence of her family, recognized her three sisters who had previously died of the same disease. Surprisingly, she then exclaimed, "and Edward too!" referring to a brother believed to be alive and well in India. The family later received letters announcing Edward's death from an accident that had occurred before his sister's vision – information that could not have reached the family by normal means.

Barrett approached these phenomena with the rigor of a scientist while remaining open to their profound implications. He noted that even Professor Charles Richet, generally skeptical of survival after death, regarded these particular deathbed visions as "impossible to explain by cryptesthesia" (a term for paranormal perception).

What distinguished Barrett's approach was his careful documentation and analysis of cases where conventional explanations seemed insufficient. When dying individuals accurately perceived deceased persons whose deaths they could not have known about through normal channels, Barrett argued this suggested something beyond hallucination or coincidence.

Barrett was especially interested in clear, lucid perceptions near death – particularly when the dying recognized deceased individuals whose deaths were unknown to them at the time. He also looked for consistent patterns across cases from different backgrounds, with an emphasis on the transformative effects on the dying person's emotional state. Barrett interpreted these visions as evidence for the existence of a spiritual world and the survival of consciousness after bodily death. While acknowledging the subjective nature of the experiences, he maintained that their consistency and evidential aspects pointed to a reality beyond material explanation. His work challenged the prevailing reductionist views of consciousness while offering comfort to those facing death or grieving loved ones. Barrett's approach balanced scientific inquiry with compassionate understanding of the human experience of mortality.

"Death-Bed Visions" established an important foundation for subsequent research into end-of-life experiences. Although published nearly a century ago, the book remains relevant to contemporary discussions in several fields such as palliative care; healthcare professionals increasingly recognize the importance of understanding and respecting deathbed experiences as part of compassionate end-of-life care. Barrett's work also raises fundamental questions about the relationship between brain function and consciousness, particularly during the dying process, in anticipation of the increasing studies of consciousness we have seen in the 21st century. In the field of thanatology, the academic study of death and dying continues to investigate phenomena similar to those Barrett documented. And lastly, the still-vibrant field of psychical research – of which Barrett is now seen as a pioneer – continues to carry his methods forward. Barrett's methodical approach to investigating anomalous experiences continues to influence researchers in the field.

Sir William Fletcher Barrett's *Death-Bed Visions* represents a significant contribution to our understanding of human experiences at the threshold between life and death. As both a distinguished physicist and a meticulous psychical researcher, Barrett brought scientific rigor to a domain often dismissed as superstition or wishful thinking.

The accounts collected in his book continue to challenge simplified materialist explanations of consciousness while offering potential insight into the perennial human question of what, if anything, awaits us beyond death. Whether interpreted as evidence of survival or as fascinating neurological phenomena, the deathbed visions Barrett documented remain worthy of serious study and consideration.

In an era of advancing medical technology that increasingly blurs the line between life and death, Barrett's work reminds us that the experience of dying may involve dimensions that extend beyond physiological processes – dimensions that deserve our attention, respect, and continued investigation.

EDITOR'S PREFACE

This book is sent out to the public in an unfinished state. It will be seen that in the introduction the author has only sketched the line of discussion which he would have elaborated, and even this outline is incomplete. The editor prefers, however, to leave untouched both the introduction and the discussion of cases, believing that by so doing the thought of the author will be more truly conveyed than would be the case were any elaboration undertaken by another hand.

The outline was clearly unfinished, for several passages in the books of reference had been marked by him for discussion, and particularly paragraphs in Professor Bozzano's recently published book, "Phenomenes Psychiques au Moment de la Mort," translated from Italian to French by C. de Vesme (Editions de la B.P.S., 8 Rue Copernic [16c], Paris), 1923.

He was specially interested in Bozzano's observation that if the phenomena were caused by the thoughts of the dying person being directed to those he loved, the appearances might be expected to represent living persons at least as frequently as deceased persons who had long passed from this world, whereas no records had come to hand of dying persons seeing at their bedside visions of friends still living.

He would have liked to ask those who believe the visions to be the product of intense desire or thought to collect evidence in support of their theory, showing that desire for living friends may produce visions of them at the bedside seen *during moments of full consciousness.*

There are, no doubt, cases of so-called traveling clairvoyance (see Chapter IV), in which the dying persons, *after a period of trance or unconsciousness,* said they had seen living relatives at a distance; and there was in some instances reciprocal vision by the distant relative of the figure of the dying person—usually mistaken for a real appearance. This is clearly a very different type of phenomenon.

Another point which the author had discussed with friends was that in the cases of phantasms of the living collected by the Society for Psychical Research it has usually been the person thinking, and not the one thought of, whose image was projected in vision.

On this analogy, when a dying person sees the phantasm of one already deceased the initiative would appear to come from the thought of the latter, whose survival is thereby demonstrated.

He was greatly impressed by a feature not uncommon at the death-beds of young children, viz., the description of the vision in terms not in keeping with ideas arising from their religious upbringing. He considered that in such circumstances the hallucination could hardly be ascribed to a mere flight of fancy.

In arranging the groups of cases he gave the first place to those relating to visions of a deceased person whom the dying percipient did not know to be dead. A recent and striking experience was that of Mrs. B., the first narrative in Chapter II. He recognized that where the death was known to anybody present in the room an attempt might be made to stretch telepathy to cover the incident, but he maintained that such an explanation would not account for the cases in this chapter, in which the percipient and the bystanders were equally unaware of the death.

The author had given considerable time and thought to the subject, and looked forward to making the groups of cases as complete and representative as possible before publication. This, however, was not to be, for he himself, in the midst of active work, passed suddenly into "that little-known country" towards which his thought had so often taken wing.

He was anxious to prove that even people who have been skeptical all their lives of any survival after death have sometimes given evidence that at the very end they knew there was an after life.

He did not therefore choose material representing visions seen only by believers in survival of the soul, or by those with special psychic powers, but also visions seen by people with no belief in a future life (see cases at the end of Chapter III).

He put each case fairly, without keeping weak points in the background, and he left it to the reader himself to consider how far telepathy or some other mental attribute could be stretched to cover the circumstances. He expected impartial critics to realize that sometimes such an explanation would appear itself to involve a flight or extension of the soul incompatible with the material bounds of life.

It is hoped that this little book, though it falls short of what the author contemplated, will to some extent carry out his plan and direct attention in this country to phenomena which seemed to him to deserve more study than they have received.

The editor gratefully acknowledges the help given by Mr. Trethewy, in his careful reading of the manuscript, in preparation of the index and in many valuable suggestions.

F. E. B.

April 1926

DEATH-BED VISIONS

CHAPTER I
INTRODUCTION

It is well known that there are many remarkable instances where a dying person, shortly before his or her transition from the earth, appears to see and recognize some deceased relatives or friends. We must, however, remember the fact that hallucinations of the dying are not very infrequent. Nevertheless, there are instances where the dying person was *unaware* of the previous death of the spirit form he sees, and is therefore astonished to find in the vision of his or her deceased relative one whom the percipient believes to be still on earth. These cases form, perhaps, one of the most cogent arguments for survival after death, as the evidential value and veridical (truth telling) character of these Visions of the Dying is greatly enhanced when the fact is undeniably established that the dying person was wholly ignorant of the decease of the person he or she so vividly sees.

With reference to these visions, the eminent physiologist of European fame, Prof. Richet, writes as follows:

> "Facts of this kind are very important. They are much more explicable on the spiritist theory than by the hypothesis of mere cryptesthesia. Among all the facts adduced to prove survival, these seem to me to be the most disquieting (i.e. from a materialistic point of view). I have therefore thought it a duty to be scrupulous in mentioning them."

As is well known, Prof. Richet does not believe in the existence of a soul or survival after death and explains the evidence afforded by psychical research of a spiritual world by his theory of *cryptesthesia*. By this, he means the perception of things or beings by some sensory organ at present unknown to science, a faculty not possessed by everyone, but, in my opinion, conclusively established to exist in certain individuals. These sensitives are to be found in all countries, in both sexes, and may be old or young, rich or poor, educated or ignorant. This faculty of clairvoyance—this vision of persons or things invisible to normal eyesight—may occur when the sensitive is quite conscious, but is more often observed in the trance condition, especially when this is induced by deep hypnosis—the "mesmeric trance" as it used to be called.

The older mesmerists employed the word "lucidity" or "travelling clairvoyance" for the perception of things at a distance. The term clairvoyance is, however, ambiguous, for it is now used in two different senses, namely, either for:

(a) The perception of hidden *material* objects remote from the sensitive, such as underground water; or

(b) The perception by the sensitive of *immaterial* objects, such as apparitions of deceased persons.

To avoid this confusion, Myers suggested the term "telesthesia" instead of clairvoyance for the perception of material things. Telesthesia he defines as the sensation or perception of objects or conditions independently of the recognized channels of sense, and also independently of any possible telepathic communication as the source of the knowledge thus gained. Hence, the term telesthesia would be inapplicable to apparitions of the dead or visions of the dying; whereas Richet would include both of these, as well as the vision of hidden material things, under his word "cryptesthesia," which appears to have the same connotation as the familiar word clairvoyance, and therefore it labours under the same ambiguity as that word.

Other terms for clairvoyance have been suggested. In America, Mr. Henry Holt uses the word "telopsis," and Dr. Heysinger the word "telecognosis"; but these terms could hardly be applied to apparitions or visions of the dying, which appear near to, and not far from, the sensitive.

Miss Cobbe, in her "Peak in Darien," makes some interesting remarks on the subject of Visions of the Dying. She states:

"The dying person is lying quietly when suddenly, in the very act of expiring, he looks up—sometimes starts up in bed—and gazes on (what appears to be) vacancy, with an expression of astonishment, sometimes developing instantly into joy, and sometimes cut short in the first emotion of solemn wonder and awe. If the dying man were to see some utterly unexpected but instantly recognized vision, causing

> him great surprise or rapturous joy, his face could not better reveal the fact. The very instant this phenomenon occurs, Death is actually taking place, and the eyes glaze even while they gaze at the unknown sight."

As regards the general subject of Visions of the Dying, Mr. Myers has some interesting remarks in "Phantasms of the Living." He states that in his view such an occurrence *"must probably often take place though it can seldom leave any record behind it. For here we have an account of that side only of a reciprocal incident which is usually lost to human knowledge altogether: I mean of the supernormal percipience of a man in the very article of death; while there is no record of any corresponding sound or vision as experienced by those to whom he seemed to pay his visit of farewell."* [1]

There are, however, several cases on record where the vision of those who have passed over is shared by friends at the bedside of the dying person. Instances of these will be given in a later chapter.

In considering the value of evidence for supernormal phenomena, the importance of the cumulative character of the evidence must be taken into account. It is the undesigned coincidence of witnesses who have had no communication with each other that constitutes its value taken as a whole, whilst a single case may be doubtful or disproved, just as a single stick may be broken but a faggot may defy all our attempts at breaking a bundle of sticks. On this point, Archbishop Whately has some admirable remarks on the value of testimony. He states:

> "It is evident that when many coincide in their testimony (where no previous concert can have taken place), the probability resulting from this concurrence does not rest on the supposed veracity of each considered separately, but on the improbability of such an agreement taking place by chance. For though in such a case each of the witnesses should be considered as unworthy of credit, and even much more likely to speak falsehood than truth, still the chances would be infinite against their all agreeing in the same falsehood."[2]

---

1 "Phantasms of the Living," Vol. II, p. 305.

2 See Whately's "Rhetoric," Chapter I

About fifty years ago, the learned incumbent of a church in Birmingham, the Rev. J. S. Pollock, published a collection of cases of supernormal phenomena under the curious title of "Dead and Gone." Although some five hundred cases are quoted, taken from various sources, no attempt has been made at the investigation of any single case, so that the book as a whole has little evidential value.

Here I may quote some suggestive remarks made by Mrs. Henry Sidgwick soon after the foundation of the Society for Psychical Research and published in the "Proceedings" for 1885 (p. 69):

> "Most of those to whom this paper is addressed probably belong to some Christian denomination, and to them the continued existence of the soul after death is, of course, no new theory invented to account for such phenomena as we are discussing, or requiring such phenomena to support it. But few will have any difficulty in agreeing with me that (1) the possibility of receiving [visions of or] communications from the dead, here and now, would not follow as a necessary consequence from the immortality of the soul; (2) that if communication of what I may call an objective kind—distinguishable, I mean, from our own thoughts and emotions—is possible to all those of the departed who desire it, we should naturally expect it to occur *more frequently* than the most sanguine can suppose that it actually does; and (3) that its possibility, while not in contradiction with any of the known facts of physical science, is certainly not supported, or in any way suggested, by any of these facts. However firmly, therefore, we may believe in the continued existence of dead human beings, we cannot regard the supposition of their action on the minds of the living as if it were merely the reference of an effect to a *vera causa* known to be adequate to produce it. We must treat it as we should treat the hypothesis—in any department of physical investigation—of an entirely new agent, for the existence of which we have no evidence outside the phenomenon which it is introduced to explain. If this be so, it will, I think, be admitted that we should be violating an established rule of scientific method if we introduced such a hypothesis except in the last resort when all other modes of explanation seem clearly to fail.
>
> "Exactly at what point of improbability this failure of other explanations is to be regarded as established cannot, I think, be defined—at any rate, I feel quite unable to define it. But I may perhaps say that, in

> my opinion, it is a point which can hardly be reached in the case of any narrative of a *single event* considered by itself: if we had only a *single* ghost-story to deal with, I can hardly conceive the kind or amount of evidence which would lead me to prefer the hypothesis of ghostly agency to all other possible explanations. The existence, therefore, of phantasms of the dead can only be established, if at all, by the accumulation of improbabilities in which we become involved by rejecting a large mass of apparently strong testimony to facts which, as recounted, would seem to admit of no other satisfactory explanation; and in testing the value of this testimony we are bound, I think, to strain to the utmost all possible suppositions of recognized causes, before we can regard the narrative in question as even tending to prove the operation of this novel agency."

On the other hand, every scientific society ought to have as its motto the opinion expressed by Sir John Herschel in his discourse on "Natural Philosophy" (p. 127), "that the perfect observer will have his eyes as it were opened that they may be struck at once with any occurrence which, according to received theories, ought *not* to happen; for these are the facts which serve as clues to new discoveries." Unfortunately, as Goethe remarked in one of his conversations with Eckermann, "in the sciences… if anyone advances anything new… people resist with all their might; they speak of the new view with contempt, as if it were not worth the trouble of even so much as an investigation or a regard; and thus a new truth may wait a long time before it can win its way."

CHAPTER II
VISIONS SEEN BY THE DYING OF PERSONS BY THEM UNKNOWN TO BE DEAD

The evidence of Visions of the Dying, when they appear to see and recognize some of their relatives of whose decease they were unaware, affords perhaps one of the strongest arguments in favor of survival. Even Prof. Richet regards this evidence as impossible to explain by cryptesthesia. I have given some striking instances of these visions of the dying in my book "On the Threshold of the Unseen," and other cases will be found in the "Proceedings" of our Society.

A recent case of the kind was related to me by Lady Barrett, which occurred when she was in attendance on a patient in the Mothers' Hospital, at Clapton, of which she is one of the Obstetric Surgeons.

Lady Barrett received an urgent message from the Resident Medical Officer, Dr. Phillips, to come to a patient, Mrs. B., who was in labor and suffering from serious heart failure. Lady Barrett went at once, and the child was delivered safely, though the mother was dying at the time. After seeing other patients, Lady Barrett went back to Mrs. B's ward, and the following conversation occurred which was written down soon afterwards. Lady Barrett says:

> "When I entered the ward Mrs. B. held out her hands to me and said, 'Thank you, thank you for what you have done for me—for bringing the baby. Is it a boy or girl?' Then holding my hand tightly, she said, 'Don't leave me, don't go away, will you?' And after a few minutes, while the House Surgeon carried out some restorative measures, she lay looking up towards the open part of the room, which was brightly lighted, and said, 'Oh, don't let it get dark—it's getting so dark... darker and darker.' Her husband and mother were sent for.
>
> "Suddenly she looked eagerly towards one part of the room, a radiant smile illuminating her whole countenance. 'Oh, lovely, lovely,' she said. I asked, 'What is lovely?' 'What I see,' she replied in low, intense tones. 'What do you see?' 'Lovely brightness—wonderful beings.' It is difficult to describe

the sense of reality conveyed by her intense absorption in the vision.

"Then—seeming to focus her attention more intently on one place for a moment—she exclaimed, almost with a kind of joyous cry, 'Why, it's Father! Oh, he's so glad I'm coming; he is so glad. It would be perfect if only W. (her husband) could come too.'

"Her baby was brought for her to see. She looked at it with interest, and then said, 'Do you think I ought to stay for baby's sake?' Then turning towards the vision again, she said, 'I can't—I can't stay; if you could see what I do, you would know I can't stay.'

"But she turned to her husband, who had come in, and said, 'You won't let baby go to anyone who won't love him, will you?' Then she gently pushed him to one side, saying, 'Let me see the lovely brightness.'

"I left shortly after, and the Matron took my place by the bedside. She lived for another hour, and appeared to have retained to the last the double consciousness of the bright forms she saw, and also of those tending her at the bedside, e.g., she arranged with the Matron that her premature baby should remain in hospital till it was strong enough to be cared for in an ordinary household.

"(Signed) Florence E. Barrett"

Dr. Phillips, who was present, after reading the above notes, writes to me saying that she "fully agrees with Lady Barrett's account."

The most important evidence is, however, given by the Matron of the Hospital, who has sent the following account:

"I was present shortly before the death of Mrs. B., together with her husband and her mother. Her husband was leaning over her and speaking to her, when pushing him aside [1] she said, 'Oh, don't hide it; it's so beautiful.' Then turning away from him towards me, I being on the other side of the bed, Mrs. B. said, 'Oh, why there's Vida,' referring to a sister of whose death three weeks previously she had not been told. Afterwards, the mother, who was present at the time, told me, as I have said, that Vida was the name of a dead sister of Mrs. B.'s, of whose illness and death she was quite ignorant, as they had carefully kept this news from Mrs. B. owing to her serious illness.

"(Signed) Miriam Castle Matron"

---

1 This is not the incident mentioned by Lady Barrett, but a later incident of the same kind.

I asked Dr. Phillips to try and obtain the independent report of Mrs. B.'s mother, who, as the Matron stated, was also present at the time. This was kindly done, and I have received the following interesting and informative letter from Mrs. Clark (Mrs. B.'s mother):

HIGHBURY, N. 5.

"I have heard you are interested in the beautiful passing of my dear daughter's spirit from this earth on the 12th day of January, 1924.

"The wonderful part of it is the history of the death of my dear daughter, Vida, who had been an invalid some years. Her death took place on the 25th day of Dec., 1923, just 2 weeks and 4 days before her younger sister, Doris, died. My daughter Doris, Mrs. B., was very ill at that time, and the Matron at the Mothers' Hospital deemed it unwise for Mrs. B. to know of her sister's death. Therefore when visiting her we put off our mourning and visited her as usual. All her letters were also kept by request until her husband had seen who they might be from before letting her see them. This precaution was taken lest outside friends might possibly allude to the recent bereavement in writing to her, unaware of the very dangerous state of her health.

"When my dear child was sinking rapidly, at first she said, 'It is all so dark; I cannot see.' A few seconds after, a beautiful radiance lit up her countenance; I know now it was the light of Heaven, and it was most beautiful to behold. My dear child said, 'Oh, it is lovely and bright; you cannot see as I can.' She fixed her eyes on one particular spot in the ward, saying, 'Oh, God, forgive me for anything I have done wrong.' After that, she said, 'I can see Father; he wants me, he is so lonely.' She spoke to her father, saying, 'I am coming,' turning at the same time to look at me, saying, 'Oh, he is so near.' On looking at the same place again, she said with rather a puzzled expression, 'He has Vida with him,' turning again to me, saying 'Vida is with him.' Then she said, 'You do want me, Dad; I am coming.' Then a very few parting words or sighs were expressed—nothing very definite or clear.

"With great difficulty and a very hard strain, she asked to see 'the man who married us': this was to her husband, who was standing on the opposite side of the bed. His name she could not say; it was the Rev. Maurice Davis, of All Saints, Haggerston, E., and he was sent for. [2] He

2 He came, but Mrs. B. had then become incapable of speech though still alive.

had known my dear child for some years, and was so impressed by the vision that he quoted it in his ' Parish Magazine ' for February last.

Yours respectfully,
"(Signed) Mary C. Clark"

Before passing on to other cases, it is desirable to discuss somewhat in detail the foregoing case. The vision seen by the dying woman, Mrs. B., was obviously not due to her normal sight; otherwise, the figures would have been seen by others present in the room. The appearance, therefore, was not due to any ordinary material objects, nor is it likely to have been due to some illusion, that is to say, the misinterpretation of some object actually present to sight—as when a dressing-gown is mistaken for a woman—for not only was there nothing in the room to suggest such an illusion, but she recognized both her deceased father and sister. Moreover, she was quite unaware of the death of the latter. A more probable explanation is that it was an hallucination, which may be defined as "a sensory perception which has no objective counterpart within the field of vision." The question, therefore, becomes whether it was merely a delusive hallucination, when there *is* nothing whatever to which it corresponds, or a veridical hallucination—corresponding to some real event, which was invisible to normal eyesight. This must not be confused with a delusion, which applies to cases where there is *no* corresponding reality. There are many well-known cases of vivid illusions of sight which sometimes accompany the oncoming of sleep, as when a dream figure persists for a short time, or when faces in the dark are vividly seen by certain persons; these illusions are termed hypnagogic. Externalized impressions of this kind are the frequent source of imaginary apparitions, such as occur to nervous people walking through lonely places at nighttime. To many of my readers, this commonsense explanation will appear to be the origin of the vision of the dying which we have just related, the whole matter being dismissed as a mere coincidence. If this case stood alone, this would be the probable explanation; it will, however, be seen that mere chance coincidence cannot apply to the numerous cases which will be recited later on. Another explanation is the creation of hallucination in the percipient by some transference of thought or telepathic influence from those around the bedside. In the case just recited, however, this explanation fails for Lady Barrett and Dr. Phillips knew nothing about the decease of the percipient's father, when the latter, looking steadily at one place, said, "Why, it's Father. Oh, he's so glad I'm coming." Nor was her husband present at the time. Moreover, the skeptical reader is likely to deny the existence of telepathy and would reject any explanation based upon that ground.

The next case has reached me from America and is a well-authenticated instance on the authority of a distinguished man, Dr. Minot J. Savage, with whom I was acquainted. Dr. Minot Savage was for many years a valued member of our S.P.R., he died in 1920. Dr. Hyslop [3] has recorded the following case in one of his books [4] and remarks: "Dr. Savage told me personally of the facts and gave me the names and addresses of the persons on whose authority he tells the incidents," which Dr. Savage narrates as follows:

> "In a neighboring city were two little girls, Jennie and Edith, one about eight years of age and the other but a little older. They were schoolmates and intimate friends. In June, 1889, both were taken ill of diphtheria. At noon on Wednesday, Jennie died. Then the parents of Edith, and her physician as well, took particular pains to keep from her the fact that her little playmate was gone. They feared the effect of the knowledge on her own condition. To prove that they succeeded and that she did not know, it may be mentioned that on Saturday, June 8th, at noon, just before she became unconscious of all that was passing about her, she selected two of her photographs to be sent to Jennie, and also told her attendants to bid her goodbye.
>
> "She died at half-past six o'clock on the evening of Saturday, June

3 As some of my readers may not be acquainted with Dr. Hyslop's name, I may mention that he was for some years the Professor of Ethics and Logic in Columbia University, New York. He studied for some years in Germany, where he took his Ph.D. and was also an LL.D. He was at first a sceptic and severe critic of psychical research, but afterwards became convinced of the importance of the subject, and resigned his university chair and all its emoluments to devote the rest of his life to the investigation of psychical phenomena. His zeal and energy and acumen were remarkable, in fact he sacrificed his life through the incessant labour involved in his duties as treasurer, hon. secretary and research officer of the American S.P.R. His literary output was enormous; he seemed to live and move and have his being in psychical research to the exclusion of almost every other subject. He spent some time with me in Ireland, and gave a learned address to the recently founded Dublin Section of the S.P.R. He died in 1920.

4 " Psychical Research and the Resurrection " (Boston, U.S.A.), 1908, p. 88.

8th. She had roused and bidden her friends goodbye, and was talking of

dying, and seemed to have no fear. She appeared to see one and another of the friends she knew were dead. So far, it was like other similar cases. But now suddenly, and with every appearance of surprise, she turned to her father and exclaimed, 'Why, papa, I am going to take Jennie with me!' Then she added, 'Why, papa! you did not tell me that Jennie was here!' And immediately she reached out her arms as if in welcome, and said, 'Oh, Jennie, I'm so glad you are here!'"

In connection with this case, Dr. Savage remarks that it is difficult to account for the incident by any ordinary theory of hallucination. If this vision were a solitary case, a mere casual coincidence might perhaps account for it, but as it is only one of a considerable group of similar cases, an explanation of chance coincidence becomes incredible. My readers will doubtless agree with Dr. Savage's remark as they peruse the other cases narrated in this volume.

The following case [5] was given in a paper contributed to the S.P.R. by Mr. Edmund Gurney and Mr. F. W. H. Myers.[6] It was received by them through the Rev. C. J. Taylor. The narrator, who does not wish his name published, was the Vicar of H:

"On November 2nd and 3rd, 1870, I lost my two eldest boys, David Edward and Harry, from scarlet fever, they being then three and four years old respectively.

"Harry died at Abbot's Langley on November 2nd, fourteen miles from my vicarage at Aspley, David the following day at Aspley. About an hour before the death of this latter child, he sat up in bed, and pointing to the bottom of the bed, said distinctly, 'There is little Harry calling to me.' Of the truth of this fact I am sure, and it was heard also by the nurse."

"(Signed) X.Z., Vicar of H-"

In letters and conversations with Mr. Podmore, Mr. Taylor adds the following details:

5 This case and the next one are quoted from pp. 99 and 100 respectively of the same book as the last. See footnote p. 8.

6 "Proceedings S.P.R.," Vol. V, p. 459.

"Mr. Z. [the Vicar] tells me that care was taken to keep David from knowing that Harry was dead, and that he feels sure that David did not know it. Mr. Z. was himself present and heard what the boy said. The boy was not delirious at the time."

The next case was communicated to the S.P.R.[7] by the Rev. J. A. Macdonald, who has for some years been a useful helper to the Society in the careful collection of evidence. Mr. Macdonald received it at first hand from Miss Ogle, who was the sister of the percipient. She writes as follows:

"My brother, John Alkin Ogle, died at Leeds, July 17th, 1879. About an hour before he expired, he saw his brother—who had died about sixteen years before—and John, looking up with fixed interest, said, 'Joe! Joe!' and immediately after exclaimed with ardent surprise, 'George Hanley!' My mother, who had come from Melbourne, a distance of about forty miles, where George Hanley resided, was astonished at this, and said, 'How strange he should see George Hanley; he died only ten days ago.' Then turning to my sister-in-law, she asked if anybody had told John of George Hanley's death; she said 'No one.' My mother was the *only* person present who was aware of the fact. I was present and witnessed this."

(Signed) Harriet H. Ogle

In answer to inquiries, Miss Ogle states:

"J. A. Ogle was neither delirious nor unconscious when he uttered the words recorded. George Hanley was an acquaintance of John A. Ogle, not a particularly familiar friend. The death of Hanley was *not* mentioned in his hearing."

The "Revue Spirite" for December, 1924, contains the following interesting case:

"The Review 'Verdade e Luz' of San Paolo, Brazil, in its number of September, 1924, has remarks on the striking incident of which the

7 See "Proceedings S.P.R.," Vol. V, p. 460.

dying Adamina Lazaro was the heroine.

"A few hours before her death, the patient said to her father that she saw near her bed several members of the family, all deceased some years previously. The father attributed this declaration *in extremis* to a state of delirium, but Adamina insisted with renewed force, and among the invisible 'visitors' named her own brother, Alfredo, who was employed at the time at a distance of 423 kilometers, on the lighthouse of the port of Sisal.

"The father was more and more convinced of the imaginary nature of these visions, well knowing that his son Alfredo was in perfect health, for a few days previously he had sent the best possible news of himself.

"Adamina died the same evening, and the next morning her father received a telegram informing him of the death of the young Alfredo. A comparison of times showed that the dying girl was still living at the time of the death of her brother."

I am indebted to Mr. C. J. Hans Hamilton for the following case, which he translated from the Review "Psychica"[8] of 1921. It was contributed by M. Warcollier, of the Institut Metapsychique, Paris, who says:

"My uncle, M. Paul Durocq, left Paris in 3 for a trip to America, with my aunt and other members of the family. While they were at Venezuela, my uncle was seized with yellow fever, and he died at Caracas on the 24th June, 1894.

"Just before his death, and while surrounded by all his family, he had a prolonged delirium, during which he called out the names of certain friends left in France, and whom he seemed to see. 'Well, well, you too—, and you—, you as well!'

"Although struck by this incident, nobody attached any extraordinary importance to these words at the time they were uttered, but they acquired later on exceptional importance when the family found, on their return to Paris, the funeral invitation cards of the persons named by my uncle before his death, and who had died before him. It is only recently that I have been able to collect the testimony of the only two survivors of this event, my cousins Germaine and Maurice Durocq."

8 Published in France.

Germaine Durocq writes as follows:

"You ask me details of the death of my poor father. I well remember him as he lay dying, though it is many years ago. The thing which probably interests you is that he told us of having seen some persons in heaven and of having spoken to them at some length. We were much astonished on returning to France to find the funeral cards of those same persons whom he had seen when dying. Maurice, who was older than I was, could give you more details on this subject."

Maurice Durocq writes:

"Concerning what you ask me with regard to the death of my father, which occurred a good many years ago, I recall that a few moments before his death my father called the name of one of his old companions—M. Etcheverry—with whom he had not kept up any connection, even by correspondence, for a long time past, crying out, 'Ah! you too,' or some similar phrase. It was only on returning home to Paris that we found the funeral card of this gentleman. Perhaps my father may have mentioned other names as well, but I do not remember."

Mr. Hans Hamilton, who translated and sent the above incident to me, comments on it as follows:

"The date of the deaths of the persons seen by M. Durocq when dying should have been verified at the time of the return of the family to Paris, since we have otherwise no *certainty* that they died before M. Durocq. However, the whole of the story makes it more than probable that this point would not have been overlooked by the family; and M. Warcollier states in his own account that the persons in question were deceased at the time of the apparitions."

The following incident was sent to the "Spectator" by "H. Wedgwood" in 1882. He says:

"Between forty and fifty years ago, a young girl, a near connection of mine, was dying of consumption. She had lain for some days in a prostrate condition, taking no notice of anything, when she opened her eyes, and looking upwards, said slowly, 'Susan—and Jane—and Ellen,' as if recognizing the presence of her three sisters, who had previously died of the same disease. Then after a short pause, she continued, 'and Edward too!'—naming a brother then supposed to be alive and well in India—as if surprised at seeing him in the company. She said no more and sank shortly afterwards. In the course of the post, letters came from India announcing the death of Edward, from an accident a week or two previous to the death of his sister.

"This was told to me by an elder sister who nursed the dying girl and was present at her bedside at the time of the apparent vision." [9]

Miss Frances Power Cobbe, Authoress of "The Peak in Darien," recites an incident of a very striking character as having occurred in a family united very closely by affection:

"A dying lady, exhibiting the aspect of joyful surprise, spoke of seeing, one after another, three of her brothers who had been long dead, and then apparently recognized last of all a fourth brother, who was believed by the bystanders to be still living in India. The coupling of his name with that of his dead brothers excited such awe and horror in the mind of one of the persons present that she rushed from the room. In due course of time, letters were received announcing the death of the brother in India, which had occurred sometime before his dying sister seemed to recognize him." [10]

Dr. E. H. Plumptre (the Dean of Wells), writing to the "Spectator," August 26, 1882, remarks:

"The mother of one of the foremost thinkers and theologians of our time was lying on her deathbed in the April of 1854. She had been for

9 See R. Pike's "Life's Borderland and Beyond," p. 29.
10 *Ibid.*, p. 18.

> some days in a state of almost complete unconsciousness. A short time before her death, the words came from her lips, 'There they are, all of them—William and Elizabeth, and Emma and Anne'; then, after a pause, 'and Priscilla too.' William was a son who had died in infancy, and whose name had never for years passed the mother's lips. Priscilla had died two days before, but her death, though known to the family, had not been reported to her." [11]

In connection with the subject of this chapter, the case of Mrs. Z. in Chapter V, p. 62, should also be read.

11 See R. Pike's " Life's Borderland and Beyond," p. 15.

CHAPTER III
VISIONS SEEN BY THE DYING OF PERSONS KNOWN BY THEM TO BE DEAD, AND DEATH-BED VISIONS SEEN BY OTHERS

"I believe no soul is left to wing its viewless flight to Paradise in solitude. I believe the 'Gloria in Excelsis' of the shining host of God welcomes the disembodied spirit upon the confines of the new world. I remember hearing once of a little dying child shrinking timidly from the idea of going alone; but just before the end, there came a spirit of sublime confidence, a supernatural opening of vision, a recognition of some companionship, and the little one cried out: 'I am not afraid; they are all here.' . . . I believe the chamber of the dying is filled with the holy angels."

- Basil Wilberforce

There are a great many records authenticated by those who have attended the last moments of a dying friend or patient, wherein shortly before death, an ecstatic vision seems to have been granted to the dying person, whose face lights up with joy and apparent recognition of some relative before he passes into the Unseen. It is needless to quote a great number of cases, as doubtless many of my readers will be familiar with instances. Such cases are not confined to one country or one nation, but they appear to be more or less common all over the world. Here, for instance, is a case that occurred amongst the Cree Indians of Saskatchewan:

The Assistant Matron of the Ahtahkakoops Indian Hospital, Sandy Lake Reserve, Saskatchewan, Canada, writes to me on January 28, 1925, about a patient in the hospital, as follows:

> "He was a Cree Indian lad, about 20 years of age, son of Chief Papewyn, of a neighbouring Reserve. He was in the last stage of phthisis and had been brought here to be cared for till the finish. He was placed in a wigwam about 100 yards distant.
>
> "At last, the supreme day arrived. It was evening, and I was with him. He was lying quietly in his bed when suddenly he sat up, stretched forth his arms with a yearning gesture, while an ecstatic smile broke

over his face. It was not simply a smile of pleasure but something far beyond it. The veil was lifted, and no one who was looking on could fail to realize that it was a glorious vision that met his gaze. He then lay back in his bed, looked at me with a smile, and passed away. He had been calm and collected during the day, there was no delirium; it was an unclouded glimpse of that higher life into which he was just entering."

"(Signed) R. Hutchinson Assistant Matron

"*Assistant Matron*"

Some interesting cases of visions seen by dying persons are given in a little book by Mrs. Joy Snell, [1] who was a nurse in a large hospital, and the cases she narrates are her own personal experiences, not narratives related at second-hand. Mrs. Snell seems to be a careful and conscientious recorder, and she has kindly furnished me with the names and other particulars of the cases given anonymously in her book.

I quote below a few of these cases as given by her:

"I recall the death of a woman (Mrs. Brown, aged 36) who was the victim of that most dreadful disease, malignant cancer. Her sufferings were excruciating, and she prayed earnestly that death might speedily come to her and end her agony. Suddenly her sufferings appeared to cease; the expression of her face, which a moment before had been distorted by pain, changed to one of radiant joy. Gazing upwards, with a glad light in her eyes, she raised her hands and exclaimed, 'Oh, mother dear, you have come to take me home. I am so glad!' And in another moment, her physical life had ceased.

"The memory of another death that occurred about the same time comes back to me. It was that of an old soldier (Mr. Auchterlonie, aged 59) who was in the last stages of tuberculosis brought on by exposure while fighting his country's battles. He was brave and patient but had frequent paroxysms of pain that were almost unendurable, and he longed for the relief which he knew death alone could bring him. One of these spasms had seized upon him, and his features were convulsed with agony as he fought for breath when he suddenly grew calm. A smile lit up his face, and looking upwards he exclaimed, with a ring of joy in his voice, 'Marion, my daughter!' Then the end came. His brother

1 "The Ministry of Angels"

and sister were at the bedside. The sister said to the brother, 'He saw Marion, his favorite daughter. She came and took him where he will suffer no more.' And she added fervently, 'Thank God! he has found rest at last.'"

In Chapter VI, other cases related by Mrs. Snell will be found.

Miss R. Canton, of Garway Road, London, W., sends me the following case, which I quote in her own words:

"Some years ago, I went to see a cousin of mine at Acton, who was very ill, and I was told by her sister that on the previous evening as she sat down on a chair by the bedside, the invalid exclaimed, 'Oh, don't J—! Oh, you have sent Mother away, she was sitting there!' and she continued to seem much distressed. My aunt had died some years previously. The dying girl told me about this herself when we were alone."

The following is a case of Vision of the Dying, translated from "La Revue Spirite" for January, 1925.

"Mr. A. R. Besancon writes as follows:

'At the commencement of February, 1915, at M-, when I was only ten years old, I had the grief of losing my mother. Her death was accompanied by circumstances which I take the liberty of relating. My mother was attended by my grandmother during her illness. One night, the latter was surprised at hearing my mother, who was sleeping in the next room, pronounce certain sentences, among others this:—"Marie, I can see you at last, I am glad you have come. Help me." (Marie was my sister who died a few years before this.) Grandmother thought it was a dream; she rose and approached my mother's bed, and to her great surprise, she found her in a perfectly normal state. My mother even told her the satisfaction she had had in seeing her daughter. Later on in the night the "conversation" was resumed, but we paid no further attention. But on the next morning, Mother was no more.

"Moreover, during the same night, one of my aunts who lived in the neighboring village of V-, had the clear impression of seeing mother. "She passed," she said to me the following day, "beside my bed without speaking, then went to embrace my two daughters and disappeared."

"Such are the facts.'"

The following case is quoted from Mr. Richard Pike's "Life's Borderland and Beyond" (p. 46):

> "In the summer of 1883, a young man named Giles, of Nottingham, had the misfortune to lose several children after long and painful periods of illness. The two eldest, Fred and Annie, aged respectively seven and eight, had died and been buried for some weeks when his little boy of four years old showed symptoms of approaching death.
>
> "The father and mother were constantly by his side, as will be readily believed, to mitigate the little fellow's sufferings as much as possible. On the night when he died, the father came to his bedside with the customary medicine, when the little boy, sitting upright in bed, cried out: 'There's Fred and Annie.' 'Where, my boy?' asked the father. 'Don't you see them there—there?' said the lad, pointing to the wall, 'they're waiting for me to go to them,' and the next minute, the little sufferer fell back on the pillow dead. It should be mentioned that the father saw nothing of the apparition to which his dying boy so vividly pointed, but he quite believes its reality."

Mrs. Kinloch, of Boundary Road, St. John's Wood, N.W., sends me instances of Visions of the Dying, which had been told her, and which I quote in her words:

> "My sister—who has recently passed over—who was with our mother when she died, told me that on the day before her death she suddenly called out, 'Oh, look at your father over there,' and pointed to a corner of the room, but my sister could see nothing.
>
> "A poor woman whom I knew told me the other day that just before her mother died, she said suddenly, 'Tom, bring the boat nearer; I can't get in.' 'Tom' was her husband."

In this case, and the next three cases, the apparitions seem to have had a more or less premonitory purpose. The incident was related to the editor of the review "Psychica," who considered it so interesting that she requested the lady to repeat it by letter, which she willingly did, only requesting that nothing more than her initials should be published, though her name and address were known to the editor of the review.

The letter is as follows:

"Dear Madam,

"With reference to the incident I related to you, which happened several years ago, the following are the facts just as they occurred:

"I lost my daughter when she was seventeen years of age; she had been ill for some five years, and for eight months before her death had been confined to her bed. During all this time, and up to her death, she maintained a remarkable degree of intelligence and will. A fortnight before her death, one evening when I was leaning over the head of her bed, I asked her what she was thinking of, seeing her absorbed. She replied, 'Little mother, look there,' pointing to the bed-curtains. I followed the direction of her hand and saw a man's form, completely white, standing out quite clearly against the dark curtain. Having no ideas of spiritism, my emotion was intense, and I closed my eyes not wishing to see any longer. My child said to me, 'You do not reply.' I had the weakness to declare to her, 'I see nothing'; but my trembling voice betrayed me doubtless, for the child added with an air of reproach, 'Oh, little mother, I have seen the same thing for the last three days at the same hour; it's my dear father who has come to fetch me.'

"My child died 15 days later, but the apparition was not repeated; perhaps it attained its greatest intensity on the day I saw it.

"(Signed) Z. G"

The editor of "Psychica" remarks:

"The lady who signs this letter is not a credulous person, and she declares that she saw the vision near the bed of her dying child at a time when her thoughts were far from the creation of a phantasmal form.

"Carita Borderieux"

(*Editor of "Psychica"*)

Mr. Hans Hamilton, who translated the above extract, remarks:

"The interest of this case lies in the fact of the apparition having taken place 15 days before death; in its being visible to two persons; and in the fact that there is not the least suspicion of either delirium or coma on the part of the dying girl."

A striking case of *collective hallucination* (that is to say, a vision seen by the relatives of the dying person as well as by the dying person herself) is given in the "Proceedings S.P.R." for 1889.[2]

The narrator, Miss Emma Pearson, writes an account of her aunt's illness and death, which is here given considerably abridged:

> "My aunt, Miss Harriet Pearson, who was taken very ill at Brighton in November, 1864, craved to be back in her own home in London, where she and her sister Ann (who had died some years previously) had spent practically all their lives. I accordingly made the necessary arrangements and had her moved home. Her two nieces (Mrs. Coppinger and Mrs. John Pearson), Eliza Quinton the housekeeper, and I did the nursing between us. She became worse and worse. On the night of Dec. 23rd, Mrs. John Pearson was sitting up with her, while Mrs. Coppinger and I lay down in the adjoining room, leaving the door ajar to hear any sound from the next room. We were neither of us asleep, and suddenly we both started up in bed, as we saw someone pass the door, wrapped up in an old shawl, having a wig with three curls on each side, and an old black cap. Mrs. Coppinger called to me, 'Emma, get up, it is old Aunt Ann!' I said, 'So it is; then Aunt Harriet will die today!' As we jumped up, Mrs. John Pearson came rushing out of Aunt Harriet's room, saying, 'That was old Aunt Ann. Where has she gone?' I said to soothe her, 'Perhaps it was Eliza come down to see how her old mistress is.' Mrs. Coppinger ran upstairs and found Eliza asleep. Every room was searched—no one was there; and from that day to this no explanation has ever been given of this appearance, except that it was old Aunt Ann come to call her sister. Aunt Harriet died at 6 p.m. that day."

Eliza Quinton, the housekeeper, confirms the above statement and adds: "We searched in every room but could not find anyone in the house. Miss Harriet died on the evening of that day, but before that she told us all that she had seen her sister, and that she had come to call her."

This last statement is further confirmed by Miss Emma Pearson in a later letter, in which she states that she remembers her Aunt saying that "her sister had come for her, for she had seen her."

In the following case, the premonitory purpose seems to be strongly marked:

2 See "Proceedings S.P.R.," Vol. VI, p. 20. Also "Human Personality,' Vol. II, p. 334.

Louise F., aged forty-eight, died after an abdominal operation in January, 1896. During her illness, she frequently asked that, when cured, she might take her little niece Lily, aged three years and three months, of whom she was very fond, to live with her in the country. About a month after the death of her aunt, little Lily, who was intelligent and precocious and in quite good health, often stopped in her play to look fixedly out of the window. Her mother asked her what she was looking at, and she answered, "It is Aunt Louise, who holds out her arms to me and calls me." Her mother, much frightened, tried to distract her attention, but the child drew her chair to the window and continued to look for several minutes. Her brother, M. F., who gave me these details, said, "I was then eleven years old and my little sister said, 'What! Don't you see Tata?' as she called her aunt. Of course, I could see nothing." For some months, nothing further was seen by the child; the visions ceased. Towards May 20th, little Lily fell ill, and when in bed she looked up to the ceiling saying that she saw her aunt calling her, surrounded by little angels. "Mother, how pretty!" she said. From day to day her illness increased, but she always repeated, "My aunt has come to fetch me; she is holding out her arms to me," and as her mother wept, she said, "Don't cry, Mother, it is very beautiful, there are angels around me." She died on the 9th of June of tubercular meningitis, four and a half months after the death of Louise F.

Such is the story told by her brother, M. F., confirmed by his sister, G. F., and her mother. The family lived very quietly in a country town. None of them know anything of psychic science.

The following case was first printed in the "Religio-Philosophical Journal," May 5, 1894. [3] Mr. B. B. Kingsbury, who contributed it, states that the informant is a member of the Presbyterian Church, and her husband confirmed her statement of voices heard by the little boy calling him. Mr. Kingsbury adds that both his informants, Mr. and Mrs. H., are worthy of the highest credit. The father is somewhat of a "sensitive," and the mother has had two or three clairvoyant experiences herself.

The statement just as it was given by the mother runs as follows:

---

3 See "Human Personality," Vol. II, p. 334.

"Had I ever doubted that there is a life beyond, my doubt would have been removed by what I call a vision. In 1883, I was the mother of two strong, healthy boys. The eldest was a bright boy of two years and seven months. The other a darling baby boy of eight months. August 6th, 1883, my baby died. Ray, my little son, was then in perfect health. Every day after baby's death (and I may safely say every hour in the day) he would say to me, 'Mamma, baby calls Ray.' He would often leave his play and come running to me, saying, 'Mamma, baby calls Ray all the time.' Every night he would waken me out of my sleep and say, 'Mamma, baby calls Ray all the time. He wants Ray to come where he is; you must not cry when Ray goes, Mamma; you must not cry, for baby wants Ray.' One day, I was sweeping the sitting-room floor, and he came running as fast as he could run, through the dining-room where stood the table with baby's high chair (which Ray now used) at the side. I never saw him so excited, and he grabbed my dress and pulled me to the dining-room door, jerked it open, saying, 'Oh, Mamma, Mamma, come quick; baby is sitting in his high chair.' As soon as he opened the door and looked at the chair, he said, 'Oh, Mamma, why didn't you hurry; now he's gone; he laughed at Ray when he passed the chair; oh, he laughed at Ray so nice. Ray is going with baby, but you must not cry, Mamma.' Ray soon became very sick. Nursing and medicine were of no avail. He died Oct. 13th, 1883, two months and seven days after baby's death. He was a child of high intelligence and matured far beyond his years. Whether it is possible for the dead to return, and whether my baby came back and was seen by his little brother or not, we leave for others to judge."

Dr. Hodgson, whose name is well known to all psychical researchers as one of the most careful and critical investigators, made inquiries regarding the case, and in reply to Dr. Hodgson's inquiries, Mrs. H. wrote:

"DEFIANCE, Ohio

"December 13th, 1894

"When the child ran to me telling me the baby was sitting in his chair at the table, there was no one in the house but the servant girl, little Ray, and myself. I told the girl nothing about it and she did not hear the child, but as soon as my husband came to dinner, I told him. After that, we talked freely of the matter to several of our friends. Little Ray knew nothing of death; we had never spoken of it to him in any way; the last time I took him to the baby's grave, shortly before he was taken sick, we were sitting by the grave, and I thought, 'Oh, if I could only take baby up and look at it for just one minute, I would feel so glad.' Instantly Ray said to me, 'Mamma, let us take baby up and look at it just one minute; then we will feel better.' Just as we were leaving the grave, he smoothed it with his little hand and said, 'Ray is going to lie down and sleep right here beside little brother, but you must not cry, Mamma.' He is now lying just where he said he would.

"P.S.—I wish to say that I have never known much of what is called modern Spiritualism, but was born and reared a Presbyterian and still belong to that Church, of which I am an active member.

"F. H."

Dr. Hodgson also wrote to Mr. H., who replied as follows:

"*Feb. 27th,* 1895

"I can truly say that my wife related it [that is, about Ray seeing the baby in the chair] to me the day it occurred when I came to dinner. I frequently heard our little boy tell his mamma that the baby called him all the time.

"W. H. H."

The following corroboration was also received by Dr. Hodgson:

"116 Summit Street
"Defiance, Ohio
"*Feb. 25th,* 1895

"Dear Sir,

"I can truly say that Mrs. and Mr. H. often spoke to me of Ray seeing the baby in the chair before he took sick. They told me the next day after it happened.

"(Mrs.) J. H. Shelters"

The following case was given by Dr. Paul Edwards, and was published in the Journal "Light" for April, 1906:

> "While living in a country town in California (U.S.A.) about the year 1887, I was called upon to visit a very dear lady friend who was very low and weak from consumption. Everyone knew that this pure and noble wife and mother was doomed to die, and at last, she herself became convinced that immediate death was inevitable, and accordingly she prepared for the event. Calling her children to her bedside, she kissed each in turn, sending them away as soon as goodbye was said. Then came the husband's turn to step up and bid farewell to a most loving wife, who was perfectly clear in her mind. She began by saying 'Newton' (for that was his Christian name) ... 'do not weep over me, for I am without pain and am wholly serene. I love you upon earth, and shall love you after I have gone. I am fully resolved to come to you if such a thing is possible, and if it is not possible I will watch you and the children from Heaven, where I will be waiting when you all come. My first desire now is to go. . . . I see people moving—all in white. The music is strangely enchanting. Oh! here is Sadie; she is with me—and—she knows who I am.' Sadie was a little girl she had lost about ten years before. 'Sissy!' said the husband, 'you are out of your mind.' 'Oh, dear! why did you call me here again?' said the wife, 'now it will be hard for me to go away again; I was so pleased while there—it was so delightful—so soothing.' In about three minutes the dying woman added, 'I am going away again and will not come back to you even if you call me.'
>
> "This scene lasted for about eight minutes, and it was very plain that the dying wife was in full view of the two worlds at the same time, for she described how the moving figures looked in the world beyond, as she directed her words to mortals in this world.
>
> "I think that of all my death scenes, this was the most impressive—the most solemn."

My friend Miss Dallas has sent me some cases of Visions of the Dying which occurred to persons she knew.

In one case, the face of her friend's mother, just before death, suddenly lighted up with intense brilliancy. When this had passed away, the dying woman opened her eyes and said that she had looked into Heaven, and had seen many people they knew who had passed over, and also that many of the things she had seen it was impossible to describe. Shortly after this, she died.

In another case, Miss Dallas tells of a widow, living with her youngest surviving son, Jim, then dying of consumption. Miss Dallas visited the mother shortly after her son's death and recorded the following in her notebook the same day:

> "Jim had died on a Thursday, and on the previous Sunday, his end appeared to be near, but he revived, and told his mother that he had seen something beautiful. Again he had a relapse, and on reviving he said he had seen two of his sisters and a brother who had died previously, but he added, 'Mother, I cannot find Bessie.' His mother told Miss Dallas that Bessie had died twelve years before when Jim was still a child. Not long after this, Jim died."

The following case is taken from the Journal of the American S.P.R. for July, 1909 (p. 422). The editor, Prof. Hyslop, relates how the original letter came into his possession, and remarks that it may be taken as documentary evidence of the incident narrated. The original letter was enclosed in one addressed to the editor of the "Open Court," a well-known American periodical. In it, the writer, Mr. William C. Church, states that the letter he forwards was written to the late Captain J. Ericsson, inventor of the Monitor, by Lady Ellen Chute, a relative of his wife, and concerns the death of Ericsson's sister-in-law, Louisa Browning. The "Amelia" referred to in the letter was the wife of Captain Ericsson, who had died in July, 1867, many years previously; and "Aunt Louisa Browning" was the sister of "Amelia."

> "Bracknell, Berks
>
> "*November 5th*, 1883
>
> "Dear Capt. Ericsson,
>
> "Since I last wrote to you our fond aunt, Louisa Browning, died on Sunday morning, October 28th, at the age of 78. On her deathbed, she appeared to see her deeply loved sister [Capt. Ericsson's wife, Amelia], who had gone before. Those watching by her heard her say—though she had before been quite unconscious—'Oh, Amelia! Amelia!' and she reached out her hand to welcome someone their earthly eyes

were not permitted to see, and then all was over. . . .

"Yours very sincerely,

"(Signed) ELLEN CHUTE"

In the case[4] here abridged, the singing and voice of the unseen visitant were heard by the mother as well as by her dying child; and a cousin of the deceased child appears to have had a vision of the child and heard a premonitory intimation of her death.

"Mrs. G., with her two little girls, Minnie and Ada, of the respective ages of eight and nine years, had been staying in the country on a visit to her sister-in-law, but having taken a house near London, she sent the two children with their nurse off by an early train, following herself by one a few hours later. Towards the evening of the same day, one of the little girls walked into the room of the house which they had quitted in the morning, where a cousin to whom she was much attached was sitting at his studies, and said to him, 'I am come to say goodbye, Walter; I shall never see you again.' Then kissing him, she vanished from the room. The young man was greatly startled and astonished, as he had himself seen both the little girls and their nurse off by the morning train.

"At this very time of the evening, both the children in London were taken suddenly ill while playing in their new home, a few hours after they had arrived. The doctor called in, pronounced their complaint to be smallpox of the most malignant kind. They both died within the week, but the youngest, Minnie, died first. The day after she was buried, the poor bereaved mother was anxiously watching the last hours of the one still left, for whom she well knew no chance of life remained. Suddenly the sick child woke up from a kind of stupor and exclaimed, 'Oh, look, Mamma, look at the beautiful angels!' pointing to the foot of the bed. Mrs. G. saw nothing but heard soft, sweet music, which seemed to float in the air. Again the child exclaimed: 'Oh, dear Mamma, there is Minnie! She has come for me'; she smiled and appeared greatly pleased. At this moment, Mrs. G. distinctly heard a voice say, 'Come, dear Ada, I am waiting for you!' The sick child smiled once again and died without a struggle."

---

4 See R. Pike's "Life's Borderland and Beyond," p, 28, in which the "Atlantic Monthly," of March, 1879 is quoted as the source.

Some time before their death, the poor mother overheard a childish conversation between the two little ones, in which the youngest, Minnie, said to the other that she felt sure she should die first, and would be certain to come and fetch her sister. This conversation was long remembered by the mother, as it was strikingly confirmed by the actual facts. It is, of course, possible that expectancy on the part of the mother (if at the time she recalled her children's conversation) may discount the evidential value of this striking case.

It has been recorded of the celebrated mathematician, Prof. De Morgan, that during the last two days of his life there were indications of his passing through the experience which he had himself considered worthy of investigation and of record. He seemed to recognize all those of his family whom he had lost—his three children, his mother and sister, whom he greeted, naming them in the reverse order to that in which they left the world. No one seeing him at that moment could doubt that what he seemed to perceive was, to him at least, visible and real. [5]

Mrs. De Morgan in her book, "From Matter to Spirit," relates the following incident, which she gives as it was told by the mother of the dying child.

> "On the morning on which John died, having bade all the family farewell, he lay for some time quite quiet, and then he spoke, his voice sounding strong and clear, and was evidently replying to some question which he had heard asked. We were astonished and awestruck. We felt that he saw and heard an angel invisible to us. Then he spoke again and said, 'Mother, here is Grandmother come! You must see her! And she is with such a great company, and they say that they are come to take me away with them.' Soon after that, he gently breathed his last." [6]

The Rev. W. G. Horder relates the following incident and says:

> "A friend of mine, of a mind naturally indisposed to faith, and at the time quite sceptical about a future life, tells me of the following incident, which made a deep impression upon him, and even wakened belief in immortality:
>
> "His brother, a young man of about 25 years of age, had be

5 See R. Pike's " Life's Borderland and Beyond," p. 15.

6 See Mrs. De Morgan's " From Matter to Spirit," p. 184.

seized with brain fever, which at last rendered him quite unconscious for about 24 hours, but just before death he raised himself in his bed, resting himself upon his hand and said, 'Who is that at the bottom of my bed?' His mother, who was sitting by his bedside, said, 'There is no one there, my dear.' He said, 'Don't you see Emma' (a departed sister) 'standing at the foot of the bed?' She said, 'No, there is no one there, my dear.' 'Yes, there is,' he said, 'it is Emma. I am coming, I am ready'; and fell back and died." [7]

The following three cases were sent to me by Mrs. Shepherd Munn, widow of the late Vicar of Orleton, Brimfield, Herefordshire, to whom all the people concerned in the narratives were known personally. She writes as follows:

"A young boy, aged fourteen, named Charles Dyer, who lived with his parents at Orleton, was dying of consumption, and had wasted away very rapidly in four or five months. During the whole of that period, he was very bright, full of interest in all around, and did not seem to be aware of his rapidly failing strength. About a week before he died, he slept in a room off his mother's, with no door between—he called her, and when she went in, he was full of excitement about a door he could see at the corner of his room, which he said was 'opening wider and wider, and when it is open wide, I shall be going through it, Mother.'

"On the morning of the day he died, his mother having left the room to fetch him something, heard him call and hastening back, found him sitting up in bed, looking towards the corner of the room, and he said to her, 'There is a nice old man come for me; he is holding out his arms for me. I must go. Don't fret, Mother'; and he fell back gently on his pillow and was gone, without any struggle for breath, and with a smile of joy on his face, which remained.

"His mother was full of ecstasy, and came down to the Vicarage that same morning to tell me of it. The impression this experience made upon her has continued to the present day, and has influenced her life for the better."

---

7 See R. Pike's "Life's Borderland and Beyond," p. 35.

The following case, also related to me by Mrs. Shepherd Munn, took place some years previous to the last, but is connected with the same family.

> "An old man, named John George—grandfather of the consumptive boy, Charles Dyer, already referred to—lay dying. He and his wife, Mary Ann George, had had a great sorrow that same year in the death of their youngest son, Tom, a young man who had been killed on the railway line on which he worked.
>
> "The dying man had been quiet for some time as though sleeping when he suddenly looked up, opened his eyes wide, and looking at the side of the bed opposite to where his wife was, exclaimed, 'Why, Mother, here is Tom, and he is all right, no marks on him. Oh, he looks fine.' Then after another silence he said, 'And here's Nance too.' A pause, then 'Mother, she is all right. She has been forgiven.' And very soon after he passed away, taking with him a sorrow which had long pressed upon the mother's heart, for Nance had fallen into sin, and had died soon after the child was born, and as the poor mother thought 'never having had time to repent.'"

The next case is also given by Mrs. Shepherd Munn, and it also, like the two preceding cases, occurred in Orleton, Herefordshire.

> "A woman, named Mary Wilding, was dying of cancer. She was passionately fond of her husband, Charles Wilding. They had worked together, brought up their children, saved some money, and bought a nice little house in Orleton, where they spent some comfortable and happy years together. When she realized that she would die and leave 'Charlie,' she became very unhappy and made them all very miserable by fretting and constantly complaining of her fate.
>
> "One day as the end drew near, when a sister of hers, who was helping to look after her, happened to be alone in the room with Mary Wilding, she suddenly looked up with such a bright expression of face and said, 'Oh, Emmie, Mother is here; she has come for me, and is going to take me with her.' She never lost the feeling of confidential joy and passed away the day after quite peacefully."

Dr. Hyslop narrates the following case, which he received from a friend whose testimony he had no reason to question:

> "I called this afternoon (May 14th, 1906) upon a lady whose child, a boy of nine years old, had died two weeks previously. He had been operated upon for appendicitis some two or three years ago and had had peritonitis at the same time. He recovered and was apparently quite well for a time. Again he was taken ill and was taken to hospital and operated upon. He was perfectly rational, recognizing his parents, the doctor and the nurse—after recovering from under the influence of the anaesthetic. Feeling that he was going, he asked his mother to hold his hands until he should be gone. Soon he looked up and said, 'Mother, dear, don't you see little sister over there?' 'No, where is she?' 'Right over there. She is looking at me.' Then the mother to pacify him said she saw the child. In a few minutes his face lighted up full of smiles, and he said, 'There comes Mrs. C.' (a lady of whom he was very fond, who had died nearly two years before), 'and she is smiling just as she used to. She is smiling and wants me to come.' In a few moments he said, 'There is Roy! I'm going to them. I don't want to leave you, but you'll come to me soon, won't you? Open the door and let them in. They are waiting outside,' and he was gone."

The mother confirms this narrative, and inquiry brings out the following facts. The "little sister" he refers to had died four years before his own birth. "Roy" is the name of a friend of the child, who had died about a year previous.

The following case is taken from the "Life of the Rev. Dwight L. Moody," the celebrated Evangelical preacher of the United States.

The last moments of Mr. Moody are described by his son, the biographer, as follows:

> "Suddenly he murmured, 'Earth recedes, heaven opens up before me. I have been beyond the gates. God is calling. Don't call me back. It is beautiful. It is like a trance. If this is death, it is sweet.'
>
> "Then his face lit up, and he said in a voice of joyful rapture, 'Dwight! Irene! I see the children's faces' (referring to his two little grandchildren, who had gone before). Turning to his wife, he said, 'Mamma, you have been a good wife to me,' and with that he became unconscious."

The following case is related by Mr. Alfred Smedley in his book of "Reminiscences" (pp. 50, 51). He gives an account of his wife's last moments and states:

> "A short time before her decease, her eyes being fixed on something that seemed to fill her with pleasant surprise, she exclaimed, 'Why! there is sister Charlotte here, and Mother and Father, and brother John and sister Mary! And now they have brought Bessie Heap!! They are all here. Oh! how beautiful! Cannot you see them?' she asked. 'No, my dear; I very much wish I could,' I answered. 'Cannot you see them?' she again asked in surprise; 'why, they are all here, and they are come to bear me away with them!' Then she added, 'Part of our family have crossed the flood, and soon the other part will be gathered home, and then we shall be a family complete in heaven!'
>
> "I may explain here that Bessie Heap had been the trusted family nurse, and my wife had always been a favorite with her.
>
> "After the above ecstatic experience my wife lingered for some time. Then fixing her gaze steadily upward again, and lifting up her hands, she joined the convoy of angel friends, who had come to usher her into that brighter spiritual world of which we had learned so little."

The next case [8] is given on the authority of Dr. Wilson of New York, who was present at the death a few years ago of the well-known American tenor, Mr. James Moore, who was a patient of his. Dr. Wilson gives the following narrative:

> "It was about four o'clock, and the dawn for which he had been watching was creeping in through the shutters, when, as I bent over the bed, I noticed that his face was quite calm and his eyes clear. The poor fellow looked up into my face, and taking my hand in both of his, he said, 'You've been a good friend to me, Doctor. You've stood by me.' Then something which I shall never forget to my dying day happened, something which is utterly indescribable. While he appeared perfectly rational and as sane as any man I have ever seen, the only way that I can express it is that he was transported into another world, and although I cannot satisfactorily explain the matter to myself, I am fully convinced that he had entered the Golden City—for he said in a stronger voice

8 See "Psychical Research and the Resurrection." J. H. Hyslop. Boston, U.S.A., 1908, p. 97.

than he had used since I had attended him, 'There is Mother! Why, Mother, have you come here to see me? No, no, I'm coming to see you. Just wait, Mother, I am almost over. I can jump it. Wait, Mother.' On his face there was a look of inexpressible happiness, and the way in which he said the words impressed me as I have never been before, and I am as firmly convinced that he saw and talked with his mother as I am that I am sitting here.

"In order to preserve what I believed to be his conversation with his mother, and also to have a record of the strangest happening of my life, I immediately wrote down every word he had said. . . . His was one of the most beautiful deaths I have ever seen."

My friend, Mrs. Carter, of St. Erth, Hayle, Cornwall, sends me the following case, which occurred on April 13, 1924, when she was present, and she wrote the following notes a few days later. She says:

"On Sunday, April 13th, I went to Hillside to sit with a Mr. Williams, who was dying of consumption, so that those belonging to him might have a little rest. He was in a state of great physical distress, and unable to lie down, and could only breathe with the greatest difficulty, with his head leaning down to within a few inches of the mattress.

"He suddenly raised himself and stretched out his hands, and said very clearly, as though speaking to someone present and whom he was glad to see, 'Edmund!! my dear brother Edmund!!' I was alone with him at the time, but when the family returned to the room later I at once related to them what he had said, and then learnt from them that his brother Edmund was dead.

"During the time that I was with him—from 3.15 to 9.15—although breathing very heavily all the time, he appeared to be quite conscious when he spoke, and called for the different members of his family. He knew me quite well, and kissed my hand and called me by my name. He also asked to have water at intervals and asked for hot tea. In spite of his great bodily distress, his trust in God remained quite unshaken, and it was very moving to hear him say at intervals, 'Dear Lord, let me go!'

"I was told that before I arrived, he had exclaimed, 'Mrs. Hooper!' She had been a great friend of his, and died here about 18 months or two years ago. He died about ten hours after I had left."

The following account of the last days of a little child was published in the "Journal of the American S.P.R.," edited by Dr. James H. Hyslop (Vol. XII, No. 6), and a considerably abridged report was compiled by Miss H. A. Dallas, [9] a summary of which is given below:

"Daisy Irene Dryden was born in Marysville, Yuba County, California, on September 9th, 1854. She died in San Jose, California, on October 8th, 1864, aged ten years and twenty-nine days.

"Her mother writes: 'In the summer of 1864 Daisy was attacked by bilious fever. After five weeks of illness, the fever left her, and for two weeks she seemed to continue to gain strength. She smiled and sang and seemed like herself again, until one afternoon, as her father sat by her bed, he noticed a singular expression on her face. It was one of both pleasure and amazement. Her eyes were directed to one place above the door. Her father asked, "Daisy, what is it? What do you see?" She replied softly, "It is a spirit, it is Jesus. And He says I am going to be one of His little lambs." "Yes, dear," said her father, "I hope you are one of His Lambs." "Oh, papa!" she exclaimed, "I am going to heaven, to Him."

"'That night she was taken with enteritis and only lived four days. She suffered much for the first twenty-four hours, being unable to retain food, water, or medicine. From that time on, she had very little pain. Her poor little body had in fact become so attenuated that there was little left for the disease to work upon. But her mind was very active and remarkably clear. Her faculties appeared sharpened. She could remember recitations she had learned in school, always having been fond of memorizing poetry. And when Lulu sang to her from the Sunday School Hymnal, she would give the name of the song and the page on which to find it.

"'She loved to have us read the Scriptures to her. I read, in John xiv, "It is expedient for you that I go away: for if I go not away the Comforter will not come unto you, but if I depart I will send Him unto you." At this, she looked up to me so heavenly as she said, "Mamma, when I go away the Comforter will come to you; and maybe He will let me come too sometimes; I'll ask Allie about it." She often said this after this time, when she felt uncertain about anything. Allie was her brother who had passed to the other life at the age of six, of scarlet fever, seven months before. He seemed to be with her a great deal of the time during those last three days because when we asked her questions which

9 See "The Nurseries of Heaven," Vale Owen and Dallas. London, 1920, p. 117.

she could not answer, she would say, "Wait till Allie comes, and I will ask him." On this occasion she waited only a short time and then said, "Allie says I may go to you sometimes; he says it is possible, but you will not know when I am there; but I can speak to your thought."

"'As I have said, Daisy lingered on the borderland for three days, after the first agonizing twenty-four hours had passed. Her physical frame had become so emaciated that there was only enough to hold the spirit in its feeble embrace; and it was manifested to us, as it were, through the thin veil of the attenuated flesh which enwrapped it. During this time, she dwelt in both worlds, as she expressed it. Two days before she left us, the Sunday School Superintendent came to see her. She talked very freely about going, and sent a message by him to the Sunday School. When he was about to leave, he said, "Well, Daisy, you will soon be over the 'dark river.'" After he had gone, she asked her father what he meant by the "dark river." He tried to explain it, but she said, "It is all a mistake; there is no river; there is no curtain; there is not even a line that separates this life from the other life." And she stretched out her little hands from the bed, and with a gesture said, "It is here and it is there; I know it is so, for I can see you all, and I see them there at the same time." We asked her to tell us something of that other world and how it looked to her, but she said, "I cannot describe it; it is so different, I could not make you understand."

"'One morning while I was in the room, putting it in order, Mrs. W., one of our kind neighbors, was reading to her these words from the Testament: "Let not your heart be troubled. In my Father's house are many mansions. I go to prepare a place for you" (John xiv, 1, 2). Daisy remarked, "Mansions, that means houses. I don't see real houses there; but there is what would be places to meet each other in. Allie speaks of going to such and such a place but says nothing of houses. You see, perhaps the Testament tells about mansions so we will feel we are going to have a home in heaven, and perhaps when I get there I'll find a home. And if I do, the heavenly flowers and trees that I love so much here—for I do see them, and that they are more beautiful than anything you could imagine—they will be there." I said, "Daisy, don't you know the Bible speaks of heaven being a beautiful city?" She said, "I do not see a city," and a puzzled look came over her face, and she said, "I do not know; I may have to go there first."

"'Mrs. W., a kind neighbor, the one who had read of the mansions to Daisy, and who was with us a great deal, told Mrs. B., a neighbor of hers, about Daisy's inner sight being open. Mrs. B. was a lady who did not believe in a future state. She was, moreover, in deep distress, having just lost her husband and a son who was about twelve years old, named Bateman. She came with Mrs. W. one evening, and, sitting beside the bed, began to ask questions. Daisy said to her: "Bateman is here, and says he is alive and well, and is in such a good place, he would not come home for anything. He says he is learning how to be good." Mrs. B. then said, "Ask him if he has seen his father." Daisy replied, "He says he has not, he is not here, and says to you, 'Mother, don't fret about me, it is better I did not grow up.'" This communication set the mother to thinking and she became a firm believer in a future state.

"'The following morning, when alone with Daisy, Mrs. W., who had brought Mrs. B. to see her, asked Daisy how she could think Mrs. B.'s son was happy. "For," said she, "when he was here, you know he was such a bad boy. Don't you remember how he used to swear and steal your playthings and break them up? You know we did not allow him to play with you nor with my children because he was so bad." Daisy replied, "Oh, Aunty, don't you know he never went to Sunday School, and was always hearing so much swearing? God knows he did not have half a chance."

"'The same day her Sunday School teacher, Mrs. H., who also was with her a great deal, was sitting beside her when Daisy said to her, "Your two children are here." Now, these children had gone to the other life several years before, and if they had lived in this world would have been nearly grown up. Daisy had never heard anyone speak of them, nor did the mother have any pictures of them, so she could not have known anything whatever about them before seeing them in the spiritual world. When asked to describe them, her description of them as full-grown did not agree with the mother's idea of them, so she said, "How can that be? They were children when they died." Daisy answered, "Allie says, 'Children do not stay children; they grow up as they do in this life.'" Mrs. H. then said, "But my little daughter Mary fell, and was so injured that she could not stand straight." To this Daisy replied, "She is all right now; she is straight and beautiful; and your son is looking so noble and happy."

"'Once she said: "Oh, papa, do you hear that? It is the singing of the angels. Why, you ought to hear it, for the room is full of it, and I can see them, there are so many; I can see them miles and miles away."

"'Mrs. W., already mentioned, who had lost her father a short time previous, wanted to know if Daisy had seen him, and brought his picture to let her see if she could recognize him. But in the evening, when she came again, Daisy told her she had not seen him, and that Allie, whom she had asked about him, had not seen him, but that Allie had said he would ask someone who could tell him about him. In a moment Daisy said, "Allie is here and says, 'Tell Aunty her father wants her to meet him in heaven, for he is there.'" Mrs. W. then said, "Daisy, why did not Allie know at once about my father?"

"Because," replied she,

"those who die go into different states or places and do not see each other at all times, but all the good are in the state of the blest."

"'During those last days of illness Daisy loved to listen to her sister Lulu as she sang for her, mostly from the Sunday School songbook. Lulu sang one song, the chorus of which was:

"'Oh! come angel band,
Come and around me stand,
Oh! bear me away on your snowy wings
To my immortal home.

"When she had finished, Daisy exclaimed, "Oh, Lulu, is it not strange? We always thought the angels had wings! But it is a mistake; they don't have any." Lulu replied, "But they must have wings, else how do they fly down from heaven?" "Oh, but they don't fly," she answered, "they just come. When I think of Allie, he is here."

"'Once I inquired, "How do you see the angels?" She replied, "I do not see them all the time; but when I do, the walls seem to go away, and I can see ever so far, and you couldn't begin to count the people; some are near, and I know them; others I have never seen before." She mentioned the name of Mary C., the sister of Mrs. S., who was a neighbor of ours in Nevada City, and said, "You know she had such a bad cough, but she is well now, and so beautiful, and she is smiling to me."

"'I was then sitting beside her bedside, her hand clasped in mine.

Looking up so wistfully to me, she said, "Dear Mamma, I do wish you could see Allie; he is standing beside you." Involuntarily I looked around, but Daisy thereupon continued, "He says you cannot see him because your spirit eyes are closed, but that I can because my body only holds my spirit, as it were, by a thread of life." I then inquired, "Does he say that now?" "Yes, just now," she answered. Then wondering how she could be conversing with her brother when I saw not the least sign of conversation, I said, "Daisy, how do you speak to Allie? I do not hear you or see your lips move." She smilingly replied, "We just talk with our think." I then asked her further, "Daisy, how does Allie appear to you? Does he seem to wear clothes?" She answered, "Oh, no, not clothes such as we wear. There seems to be about him a white, beautiful something, so fine and thin and glistening, and oh, so white, and yet there is not a fold, or a sign of a thread in it, so it cannot be cloth. But makes him look so lovely." Her father then quoted from the Psalmist: "He is clothed with light as a garment." "Oh, yes, that's it," she replied.

"'She often spoke of dying, and seemed to have such a vivid sense of her future life and happiness that the dread of death was all dispelled. The mystery of the soul's departure was to her no more a mystery. It was only a continuation of life, a growing up from the Conditions of earth-life into the air and sunshine of heaven.

"'The morning of the day she died, she asked me to let her have a small mirror. I hesitated, thinking the sight of her emaciated face would be a shock to her. But her father, sitting by her, remarked, "Let her look at her poor little face if she wants to." So I gave it to her. Taking the glass in her two hands, she looked at her image for a time, calmly and sadly. At length she said, "This body of mine is about worn out. It is like that old dress of Mamma's hanging there in the closet. She doesn't wear it anymore, and I won't wear my body anymore, because I have a new spiritual body which will take its place. Indeed, I have it now, for it is with my spiritual eyes I see the heavenly world while my body is still here. You will lay my body in the grave because I will not need it again. It was made for my life here, and now my life here is at an end, and this poor body will be laid away, and I shall have a beautiful body like Allie's." Then she said to me, "Mamma, open the shutters and let me look out at the world for the last time. Before another morning, I shall be gone." As I obeyed her loving request, she said to her father, "Raise me up, Papa." Then, supported by her father, she looked through the window whose shutters I had opened and called out, "Goodbye, sky. Goodbye, trees. Goodbye, flowers. Goodbye, white rose. Goodbye, red rose. Goodbye, beautiful world," and added, "how I love it, but I do not wish to stay."

"That evening, when it was half-past eight, she herself observed the time, and remarked, "It is half-past eight now; when it is half-past eleven Allie will come for me." She was then, for the time being, reclining on her father's breast, with her head upon his shoulder. This was a favorite position, as it rested her. She said, "Papa, I want to die here. When the time comes, I will tell you."

"'Lulu had been singing for her, and as half-past eight was Lulu's bedtime she arose to go. Bending over Daisy, as she always did, she kissed her, and said, "Good night." Daisy put up her hand and, stroking tenderly her sister's face, said to her, "Good night." When Lulu was half-way up the stairs, Daisy again called out after her, in a clear, sweet, earnest tone, "Good night and goodbye, my sweet, darling Lulu."

"'At about a quarter past eleven she said, "Now, Papa, take me up; Allie has come for me." After her father had taken her, she asked us to sing. Presently someone said, "Call Lulu," but Daisy answered promptly, "Don't disturb her, she is asleep," and then, just as the hands of the clock pointed to the half-hour past eleven, the time she had predicted that Allie was to come to take her with him, she lifted up both arms and said, "Come, Allie," and breathed no more. Then tenderly laying her loved but lifeless form upon the pillow, her father said, "The dear child has gone," and added, "she will suffer no more."'

There are one or two specially interesting points about this case—like Case 1 in Chapter II—the dying child kept a consciousness of the visions which came to her, together with clear recognition of her earthly friends, and the ability to converse with them sensibly. With Daisy Dryden, the double consciousness lasted a few days, whereas in the case of Mrs. B., it only lasted an hour or two.

Again, the descriptions Daisy gave of her vision evidently did not accord with her preconceived ideas of a spiritual world, yet it did not once occur to her to doubt the reality of what she was learning of a life apart from a material body—and the possession of a spiritual body.

In p. 118 of "The Nurseries of Heaven" (see p. 56 above) is the following statement made by her mother: "Although she was on the whole a good child, possessing ordinary good sense, yet in no way was she more remarkable than many other children. Her dying experience, therefore, was not the outgrowth of a life highly spiritual, nor was it one which had been educated in the least degree on the lines of mysticism or modern spiritualism."

Her father was so deeply impressed "by what she most undoubtedly said, heard and revealed to them," that he began a careful study of the New Testament in the original Greek, and published a series of articles later on the subject.[10]

The following incident, taken from the "American S.P.R. Journal" for 1918 (Vol XII, p. 623), was reported by Dr. E. H. Pratt, of Chicago:

> "My sister Hattie, while attending school at Mt. Carroll Seminary, suffered an attack of malignant diphtheria. She was brought home to be under our father's care, but he was unable to save her, and after a few days of extreme suffering, her spirit took its flight into what seems to most of us such a dark, impenetrable expanse of appalling immensity. A deathbed scene occurred, so wonderful, realistic, and impressive, that although I was but ten years of age at the time, my memory picture of that event is as vivid and distinct as though it were taken but yesterday.
>
> "Her bed was in the middle of the living room, and my mother, father, younger sister, and a few friends were standing about it, gazing earnestly upon my sister's dear features, as the light of life gradually went out, and the ashy pallor of death settled over them. Hattie's going out was not abrupt. It was a gradual fading away, very calm and apparently free from pain. Although her throat was so choked up with diphtheritic membrane that her voice was very thick, and it required close attention to catch all of her words, her mind seemed unusually clear and rational.
>
> "She knew she was passing away and was telling our mother how to dispose of her little personal belongings among her close friends and playmates when she suddenly raised her eyes as though gazing at the ceiling toward the farther side of the room, and after looking steadily and apparently listening for a short time, slightly bowed her head, and said, 'Yes, Grandma, I am coming, only wait just a little while, please.' Our father asked her, 'Hattie, do you see your grandma?' Seemingly surprised at the question she promptly answered, 'Yes, Papa, can't you

10 "Resurrection of the Dead," published by Hitchcock and Walden, Cincinnati, in 1872.

see her? She is right there waiting for me.' At the same time, she pointed toward the ceiling in the direction in which she had been gazing. Again, addressing the vision she evidently had of her grandmother, she scowled a little impatiently and said, 'Yes, Grandma, I'm coming, but wait a minute, please.' She then turned once more to her mother and finished telling her what of her personal treasures to give to different ones of her acquaintances. At last giving her attention once more to her grandma, who was apparently urging her to come at once, she bade each of us goodbye. Her voice was very feeble and faint, but the look in her eyes as she glanced briefly at each one of us was as lifelike and intelligent as it could be. She then fixed her eyes steadily on her vision but so faintly that we could just catch her words, said, 'Yes, Grandma, I'm coming now.' Then without a struggle or evidence of pain of any kind, she gazed steadily in the direction she had pointed out to us where she saw her grandma, until the absence of oxygen in her bloodstream, because respiration had ceased, left her hands and face all covered with the pallor of lifeless flesh.

"She was so clear-headed, so positive of the vision and presence of her grandma, with whom she talked so naturally, so surprised that the rest of us could not see grandma, the alternation of her attention and conversation between her grandma and father and mother were so distinctly photographed upon the camera of my brain that I have never since been able to question the evidence of the continuance of distinct recognizable life after death. Her grandmother had died a few years previously, and before that she and grandma had always been such close friends, and the recognition of each other as Hattie left her body to join her dearly beloved grandma in the realms beyond the vision of our physical eyes was so unquestionable and complete in every detail that it seems impossible to account for the remarkable event on any theory except that her grandma was alive and so completely like herself while on earth that Hattie's recognition of her was instantaneous and unquestionable, a real genuine experience."

The following case was communicated to the American S.P.R. by Mr. S. B. Bennett (see "American S.P.R. Journal" for 1918, Vol. XII, p. 607):

"Pittston, Pa.,

"*December* 15, 1906

"Mr. G. H. Tench died in 1902, after years of patient though intense suffering of cancer. He lived in Wilkes-Barre, but was formerly a near neighbor of mine in West Pittston, during a portion of the time he was a foreman under me enjoying mutual confidence and esteem. He received deserved promotion by another Coal Co., but our personal relationship remained the same.

"During the last weeks, I watched with him as often as I could, going back and forth by rail. While suffering intensely he would not take narcotics nor stimulating medicine, saying, 'I have lived Hall Tench and I am going to die that way.' The night the end came, he roused his younger son, telling him to call the family as he was going away. He talked entirely rationally to them and was fully conscious. Later a brother came to the house and upon entering the room G. H. Tench said, 'Goodbye, Will; I am going soon,' and closed his eyes. The family thought the end had come, but after a short interval he opened his eyes and, looking over and above the bed foot, with raised head and every appearance of interest, said clearly and distinctly, 'Why, they're all plain people.' This closed the scene, which was described to me by his wife soon after the funeral.

"Now Tench was not a religious man, although attended by a Methodist minister at the last, but a moral, upright man in every relation of life, thoroughly courageous, as was shown by his refusal to have his sensibilities dulled in his suffering. Not highly educated, nor a great reader, yet I have no doubt he had thought about conditions he had to face, and was *likely* to have imbibed the wings and harp idea. Is it not possible that he at the last expressed surprise that the people waiting for him should be 'all plain people'? I give you this as a *fact*,

"(Signed) S. B. Bennett"

The following narrative was recorded in the "Journal of the American Society for Psychical Research" (1918, p. 603), having been sent to Prof. Hyslop by Mr. Rud. C. Gittermann, a member of the English S.P.R. He writes as follows:

"My father died in Germany on March 18th, 1892, and my mother then came to live with us at Odessa. Shortly after, she fell ill and died on May 6th of the following year, 1893. Both she and my father had always been most skeptical of anything concerning the existence and survival of the soul.

"A few minutes before her death she regained consciousness (having been in a state of coma for two hours previously), raised herself in her bed, stretched out her arms, and with a happy smile on her face, cried out, 'Papa! Papa!' just as if she suddenly saw him in front of her. Immediately after she fell back into the arms of my wife and expired.

"My mother used to call her husband 'Papa,' just as we children did.

"I certify that this is a perfectly true account of what took place.

"(Signed) RUD. C. GITTERMANN"

The following abridged account of the last days of the American poet, Horace Traubel, is taken from a fuller narrative in the American S.P.R. "Journal" for 1921 (Vol. XV, pp. 114-123).

Horace Traubel (1858-1919) was the Boswell of Walt Whitman; he was also the author of a number of volumes of poems of the Whitman type, which some of his disciples regard as equaling those of his master. He was also the founder of the well-known Contemporary Club of Philadelphia.

The abridged account was contributed by Mrs. Flora Macdonald Denison, who was present at the deathbed, to the April-May issue of a Magazine entitled, "The Sunset of Bon Echo," as follows:

"All day on August 28th Horace was very low-spirited. Anne's illness and the going of the Bains were too much for him. Mildred was with him a good deal, and we decided not to leave him a minute. He had been brought in from the veranda but absolutely radiant, and on seeing me, he called out, 'Look, look, Flora, quick, quick, he is going.' 'What, Horace,' I said, 'what do you see? I cannot see anyone.' 'Why just over the rock, Walt appeared, head and shoulders and hat on in a golden glory—brilliant and splendid. He reassured me—beckoned to me, and spoke to me, I heard his voice but did not understand all he said, only "Come on."'

"Frank Bain soon came in and he repeated the story to him. All the rest of the evening Horace was uplifted and happy. So often Horace would say, 'Do not despise me for my weakness,' but now he was quite confident, even jocular, as I handed him a drink.

"On the night of September 3rd Horace was very low. I stayed for a few hours with him. Once his eyes rolled; I thought he was dying, but he just wanted me to turn him. As I did so, he listened and seemed to hear something. Then he said, 'I hear Walt's voice, he is talking to me.' I said, 'What does he say?' He said, 'Walt says, "Come on, come on."' After a time, he said, 'Flora, I see them all about me, Bob and Bucke and Walt and the rest.'

"Colonel Cosgrave had been with Horace in the afternoon and had seen Walt on the opposite side of the bed, and felt his presence. Then Walt passed through the bed and touched the Colonel's hand, which was in his pocket. The contact was like an electric shock. Horace was also aware of Walt's visible presence and said so. There was no gloom about the house. No one seemed depressed. A feeling of triumph, of pride, and of exultation permeated the atmosphere."

A letter was afterward received by Mr. Walter Prince of the American S.P.R., from Col. Cosgrave, confirming the statement given by Mrs. Flora Denison as above.

There are several cases of which records have been preserved in the "Proceedings" of the Society for Psychical Research and elsewhere, in which an account is given of those watching beside a dying relative having had a vision of spirit forms near the bed.

"In one case, [11] two women watching by their dying sister, Charlotte, saw a bright light and within it two young faces hovering over the bed, gazing intently at Charlotte; the elder sister recognized these faces as being two of her brothers, William and John, who had died when she was young. The two sisters continued to watch the faces till they gradually 'faded away like a washed-out picture,' and shortly afterwards their sister Charlotte died."

Mr. Podmore, who investigated this case, remarks that it is possible the vision was due to a telepathic impact from the dying person, but this explanation is less tenable and quite as unlikely as is the percipience of spirit forms by the dying person and sometimes

11 See "Proceedings S.P.R.," Vol. VI, p. 293.

by those present. Mr. Podmore, with his usual prejudice against any supernormal explanation remarks on this that "the images traditionally associated with death receive a sensory embodiment," but this point of view is not appropriate to the two following cases where the percipients being very young children could hardly be supposed to have any mental images traditionally associated with death, nor does it account for the "collective hallucination" described in Miss Pearson's case, pp. 35, 36.

This case is quoted by Stainton Moses:

> "Miss H., the daughter of an English clergyman, was tending a dying child. His little brother, aged three to four years, was in a bed in the same room. As the former was dying, the little brother woke up, and, pointing to the ceiling with every expression of joy, said, 'Mother, look at the beautiful ladies round my brother! How lovely they are, they want to take him.' The child died at that moment."

Another instance is reported by M. Pelusi, librarian at the Victor Emmanuel Library at Rome (Luce e Ombra, 1920, 20):

> "A little girl of three, Hippolyte Notari, partly paralyzed, was in the same room with her little brother of four months, who was dying. The father, the mother, and the grandmother of the two children were present. About fifteen minutes before the death of the infant, little Hippolyte stretched out her arms, saying, 'Look, mother, Aunt Olga.' This Aunt Olga was a younger sister of Mme. Notari, who had killed herself a year previously owing to a disappointment in love. The parents asked, 'Where do you see Aunt Olga?' The child said, 'There, there!' and tried insistently to get out of bed to go to her aunt. They let her get up, she ran to an empty chair, and was much discountenanced because the vision had moved to another part of the room. The child turned around and said, pointing to a corner, 'Aunt Olga is there.' Then she became quiet and the baby died."

In the following case which was communicated by Prof. W. C. Crosby, Associate Member, Society for Psychical Research ["Proceedings S.P.R.," Vol. VIII, pp. 229-231] the vision was seen by the nurse during the unconsciousness of the dying patient. The phantom seen was unknown to the percipient.

"Mrs. Caroline Rogers, seventy-two years old, a widow who had been twice married, and whose first husband, a Mr. Tisdale, died about thirty-five years ago, has lived on Ashland Street, in Roslindale, Mass., for the last twenty-five years; and since the death of her last child some years ago she has lived quite alone. Early in March of this year she was stricken with paralysis, and after an illness of nearly six weeks died on the afternoon of Tuesday, April 15th.

"Mrs. Mary Wilson, a professional nurse, forty-five years old, attended Mrs. Rogers during her illness, remaining with her almost constantly until she died. She had never seen Mrs. Rogers before the latter's illness, and knew nothing of her family or history. Mrs. Rogers spoke frequently to Mrs. Wilson, and also to others, as had long been her custom, of her second husband, Mr. Rogers, and children, expressing a desire to see them again, etc.

"On the afternoon of April 14th, Mrs. Rogers became unconscious, and remained so all the time until her death twenty-four hours later. Mrs. Wilson sat up with her through the whole of Monday night. Mrs. Wilson's daughter Ida, twenty-five years old, kept her mother company, and a boy of ten or twelve years slept in an adjoining chamber, to be called in case of an emergency. These four were the only persons in the house. The outer doors were securely locked, the door leading from the sick chamber on the second floor into the hall was kept locked all the time because it was near the foot of Mrs. Rogers' bed; and entrance to the sick chamber was gained by passing from the upper hall into the living room by a door which was locked that night, and thence through the chamber in which the boy slept—the two chambers having been made to communicate by cutting a door through the back of a small closet. This door was diagonally facing the bed on which Mrs. Rogers lay. Mrs. Wilson rested on a settee placed at right angles to the head of Mrs. R/s bed, so that when lying down her face was almost directly opposite this door and not more than ten or twelve feet from it. The lamp, which burned brightly all night, stood on a small table in the corner of the room directly opposite the door; and Ida occupied a couch against the wall and between the lamp and door.

"Mrs. Wilson was pretty well worn out with her long vigil;

believing that Mrs. Rogers was dying, she was naturally very nervous and timid; and having heard Mrs. R. speak frequently of seeing her departed friends, etc., she had a feeling of expectancy and dread with regard to supernatural visitations. Between two and three a.m., while her daughter was asleep, and while she was resting on the settee, but wide awake, she happened to look toward the door into the adjoining chamber and saw a man standing exactly in the doorway, the door being kept open all the time. He was middle-sized, broad-shouldered, with shoulders thrown back, had a florid complexion, reddish-brown hair (bareheaded) and beard, and wore a brown sack overcoat, which was unbuttoned. His expression was grave, neither stern nor pleasant, and he seemed to look straight at Mrs. Wilson, and then at Mrs. Rogers without moving. Mrs. Wilson supposed, of course, that it was a real man, tried to think how he could have got into the house. Then, as he remained quite motionless, she began to realize that it was something uncanny, and becoming frightened, turned her head away and called her daughter, who was still asleep on the couch, awakening her. On looking back at the door after an interval of a minute or two the apparition had disappeared; both its coming and going were noiseless, and Mrs. Rogers remained perfectly quiet, and so far as could be known entirely unconscious during this time. The chamber into which this door leads being quite dark, there was no opportunity to observe whether or not the apparition was transparent. Mrs. Wilson shortly afterwards went into this chamber and the living room, but did not examine the lower part of the house until morning, when the doors were found properly locked and everything all right.

"In the morning Mrs. Rogers' niece, Mrs. Hildreth, who lives in the neighborhood, and has known Mrs. R. and her family for many years, called at the house. Mrs. Wilson related her experience to her and asked if the apparition resembled Mr. Rogers, and Mrs. Hildreth replied emphatically that it did not. (All who knew Mr. Rogers are agreed on this point.) Their conversation was interrupted then, but when resumed later in the day Mrs. Hildreth said that Mrs. Wilson's description agreed exactly with Mr. Tidsale, Mrs. Rogers' first husband. Mrs. Rogers came to Roslindale after marrying Mr. Rogers, and Mrs. Hildreth is the only person in that vicinity who ever saw Mr. Tisdale; and in Mrs. Rogers' house, there is no portrait of him nor anything suggestive of his personal appearance. Mrs. Wilson is also very positive that the apparition was unlike anyone she ever knew.

"Mrs. Wilson has had similar experiences before, and at least one, which occurred when she was eighteen years old, which appears to have been veridical.

"The foregoing account of my experience is correct in every particular.

"(Signed) Mary Wilson"

"The foregoing is a full and accurate statement of Mrs. Wilson's experience as she related it to me on the morning of April 15th.

"(Signed) F. E. Hildreth"

June 5th, 1890

"Mrs. Wilson and Mrs. Hildreth have both impressed me as being intelligent and perfectly honest and truthful; and I have no doubt that Mrs. Wilson's experience was real and substantially as she has described it.

"(Signed) W. O. Crosby"

CHAPTER IV
VISIONS SEEN BY THE DYING OF LIVING PERSONS AT A DISTANCE - IN SOME CASES RECIPROCAL

We now come to a somewhat different and large class of cases where the veil that hides the spiritual world is not lifted for a few moments for the dying percipients, but their souls appear to be transported to a different place *on Earth*, allowing them to see persons who may be at a remote distance. Such cases are usually called instances of "traveling clairvoyance," and numerous well-attested facts of this kind have been collected in "Phantasms of the Living," to which my readers are referred.

There are, however, a few cases worthy of special notice, wherein the dying persons appear not only to make themselves visible at a distance but also inform those around them where they have been and that they have visited those whom they desired to see.

One of the most remarkable and pathetic of these so-called "reciprocal cases" was related to me by that gifted and venerable Quaker lady, Miss Anna Maria Fox, when we were on a voyage to Canada for the British Association Meeting in 1884. Miss Fox and her sister Caroline were well-known to *savants* in the last generation, [1] for their beautiful place "Penjerrick," near Falmouth in Cornwall, was the rendezvous of many eminent scientific and literary men, and nearly fifty years ago I had the privilege of enjoying their hospitality. When narrating the incident, Miss Fox referred me to her relatives, the Birkbecks, for confirmation of it: and this was given me when I made inquiries shortly afterward.

Mr. Myers has given an abridged record of the same case, [2] which he obtained from another member of the same family, Mrs. Charles Fox of Falmouth, who had heard the account from one of the percipients. The incident is nearly two centuries old, but as Mr. Myers says, the Fox family is one which would carefully preserve evidence of

1 See "Memoirs of Caroline Fox."

2 See "Phantasms of the Living," Vol. II, p. 560.

this kind. As an illustration of this fact, I may state that the narrative which Miss Anna Maria Fox gave me was practically identical with that given by Mrs. Charles Fox, which I now quote:

> "In 1739, Mrs. Birkbeck, wife of William Birkbeck, banker, of Settle, and a member of the Society of Friends, was taken ill and died at Cockermouth, while returning from a journey to Scotland, which she had undertaken alone—her husband and three children, aged seven, five, and four years respectively, remaining at Settle. The friends at whose house the death occurred made notes of every circumstance attending Mrs. Birkbeck's last hours, so that the accuracy of the several statements as to time, as well as place, was beyond the doubtfulness of man's memory or of any even unconscious attempt to bring them into agreement with each other.
>
> "One morning, between seven and eight o'clock, the relation to whom the care of the children at Settle had been entrusted, and who kept a minute journal of all that concerned them, went into their bedroom as usual, and found them all sitting up in their beds in great excitement and delight. 'Mamma has been here!' they cried, and the little one said, 'She called "Come, Esther!"' Nothing could make them doubt the fact, and it was carefully noted down, to entertain the mother on her return home. That same morning, as their mother lay on her dying bed at Cockermouth, she said, 'I should be ready to go if I could but see my children.' She then closed her eyes, to reopen them, as they thought, no more. But after ten minutes of perfect stillness she looked up brightly and said, 'I am ready now; I have been with my children,' and then at once peacefully passed away. When the notes taken at the two places were compared, the day, hour, and minutes were the same.
>
> "One of the three children was my grandmother, *née* Sarah Birkbeck, afterward the wife of Dr. Fell, of Ulverston. From her lips I heard the above almost literally as I have repeated it. The eldest was Morris Birkbeck, afterward of Guildford. Both these lived to old age and retained to the last so solemn and reverential a remembrance of the circumstance that they rarely would speak of it. Esther, the youngest, died soon after. Her brother and sister heard the child say that her mother called her, but could not speak with any certainty of having themselves heard the words, nor were they sensible of more than their mother's standing there and looking on them."

The case of Mrs. Goffe is also of remote date, 1691, but is taken from a contemporary report made by the Rev. T. Tilson in a letter he addressed to the famous divine, Richard Baxter, who published it in a book he wrote. [3] The case is given in "Phantasms of the Living" (Vol. II, pp. 558-559), and the authors state that the narrative cannot be impugned on the ground of any credulity on the part of Baxter, and quote an authority on this point. It will be seen that the incidents in the following narrative are curiously parallel to the preceding case of Mrs. Birkbeck. Though Mr. Tilson's letter which we now quote is somewhat long, it is better to give his own words rather than an abstract.

"*July 6th*, 1691

"Mary, the wife of John Goffe, of Rochester, being afflicted with a long illness, removed to her father's house at West Mulling, which was about nine miles distant from her own; there she died, June 4th, 1691. The day before her departure she grew impatiently desirous to see her two children, whom she had left at home, to the care of a nurse. She prayed her husband to hire a horse, for she must go home to die with her children. Between one and two o'clock in the morning she fell into a trance. One widow Turner, who watched with her that night, says that her eyes were open and fixed, and her jaw fallen; she put her hand on her mouth and nostrils but could perceive no breath; she thought her to be in a fit, and doubted whether she was alive or dead. The next day this dying woman told her mother that she had been at home with her children. 'That is impossible,' said the mother, 'for you have been here in bed all the while.' 'Yes,' replied the other, 'but I was with them last night while I was asleep.'

"The nurse at Rochester, widow Alexander by name, affirms and says she will take her oath of it before a magistrate, and receive the sacrament upon it, that a little before two o'clock that morning she saw the likeness of the said Mary Goffe come out of the next chamber (where the elder child lay in a bed by itself, the door being left open), and stood by her bedside for about a quarter of an hour; the younger child was there lying by her; her eyes moved, and her mouth went, but she said nothing. The nurse, moreover, says that she was perfectly awake; it was then daylight, being one of the longest days in the year. She sat up in her bed and looked steadfastly upon the apparition; at that time

3 See Baxter's "The World of Spirits" (1691), pp. 147-51

she heard the bridge clock strike two, and a while after said, 'In the name of the Father, Son, and Holy Ghost, what art thou?' Thereupon the appearance removed and went away; she slipped on her clothes and followed, but what became of it she cannot tell. Then, and not before, she began to be grievously affrighted, and went out of doors, and walked upon the wharf (the house is just by the riverside) for some hours, only going in now and then to look at the children. At five o'clock she went to a neighbor's and knocked at the door, but they would not rise; at six she went again, then they rose and let her in. She related to them all that had passed; they would persuade her she was mistaken, or dreamt; but she confidently affirmed, 'If ever I saw her in all my life, I saw her this night.'" [The writer then gives an account of how one of those to whom she related the story confirmed the above narrative.]

"The substance of this statement was related to me by John Carpenter, the father of the deceased, the next day after the burial—July 2. I fully discoursed the matter with the nurse and two neighbors, to whose house she went that morning. Two days after I had it from the mother, the minister that was with her in the evening, and the woman who sat up with her last that night. They all agree in the same story, and every one helps to strengthen the other's testimony. They all appear to be sober, intelligent persons, far enough off from designing to impose a cheat upon the world, or to manage a lie; and what temptation they should lie under for so doing I cannot conceive.

"(Signed) Thomas Tilson"

The next case, also contributed by Mr. Myers, is an account given by the Ellis family to Mr. Myers, of a vision which their father, Mr. Ellis, who was dying in Kensington, had of his son, Robert, at that time in Australia. The Misses Ellis state:

"On Wednesday, December 29th, 1869, my father, who was dangerously ill at the time, awoke from a sleep, and raising himself up in the bed pointed and looked most intently to one corner of the room and said to us (my sister Mary and me), 'Look! don't you see? it is my poor boy Bob's head!' Then turning to me, he said, 'Norman Town, don't forget, Gulf of Carpentaria.' He then sank back exhausted. This happened about three p.m. We found, after his death, he had entered the address in red ink in his pocketbook—my brother having left Bourke Town and gone to Norman Town—so that the next packet of letters was sent there. My father died on Thursday, Dec. 30th, 1869. When my brother returned from Australia a few years after, he told us that one night, whilst camping out, he had gone to rest and had slept, and he awoke seeing my father's head distinctly in one part of his tent. It made such an impression on him that he went to his mate in the adjoining tent and said, 'I have seen my father; you must come and stay with me.' By the next mail he received my letter telling him of my"

"My brother said it must have been about three a.m. when he saw my father. Would not that correspond with our three p.m.? I always think they must have seen each other at the same time.

"(Signed) Alice Ellis
"Mary Ellis"

Mr. Myers states that in conversation with the narrators, he ascertained that Mr. Ellis was not in the least delirious during his last days, and that he was deeply attached to his absent son.

In this case in connection with the Vision of his father seen by Mr. Robert Ellis, it may be interesting to note that another case of apparition, occurring to her husband some years later, is given by Mrs. Robert Ellis. She states that on *Tuesday, December* 19, 1876, between 6 and 7 p.m., when she and Mr. Ellis were sitting talking together, he suddenly looked over his shoulder with a startled, almost terrified look. On being asked what was the matter, he said that he had imagined he saw someone coming in at the door. Subsequently, he stated that he distinctly saw the tall dark figure of a man, but could not distinguish his features. He was greatly agitated. Later on, a telegram was received, giving news of the sudden death of Mrs. Robert Ellis' brother in Mexico on *Tuesday*, the *19th December*, at seven o'clock in the evening. He and Mr. Robert Ellis had been very great friends.

This case is taken from "Phantasms of the Living," Vol. II, p. 253:

"The lady who sends us the following narrative occupies a position of great responsibility and desires that her name may not be published, but it may be given to inquirers:

"'When I was eight months old, my mother's youngest sister, Mercy Cox, came to reside with us and to take charge of me. My father's position at the Belgian Court as portrait painter obliged him to be much abroad, and I was left almost wholly to the care of my very beautiful aunt. The affection that subsisted between us amounted almost to idolatry, and my poor mother wept many bitter tears when she came home to see how little I cared for anyone else. My aunt took cold, and for three years lingered in decline. I was a quick child and could read well and even play prettily, so that I was her constant companion day and night. Our doctor, Mr. Field, of the Charter House, greatly disapproved of this close contact, and urged my parents to send me quite away. This was a difficult feat to accomplish, the bare mention of the thing throwing my aunt into faintings. At last Mr. Cumberland (the theatrical publisher) suggested that I should join his two daughters, Caroline, aged 16, and Lavinia, younger, at Mrs. Hewetson's, the widow of a clergyman resident at Stourpaine, in Dorsetshire, who only took four young ladies. This was represented to my aunt as something so wonderfully nice and advantageous to me that she consented to part with me. My portrait was painted and placed by her bed, and I remember how constantly she talked to me about our separation. She knew she would be dead before the year of my absence would be ended. She talked to me of this and of how soon I should forget her; but she vehemently protested that she would come to me there. Sometimes it was to be as an apple-woman for me to buy fruit from, sometimes as a maid wanting a place, always *she* would know *me*, but I should not know her until I cried and implored to know her.

"I was but nine when they sent me away, and coach traveling was very slow in those days. Letters too were dear, and I very rarely had one. My parents had sickness and troubles, and they believed the reports that I was well and happy, but I was a very miserable, ill-treated little girl. One morning at break of day—it was New Year's Day—I was sleeping beside Lavinia. We two shared one little white tester bed with curtains, while Caroline—upon whom I looked with awe, she being 16—slept in another similar bed at the other end of a long, narrow room, the beds being placed so that the feet faced each other, and two white curtains hung down at the sides of the head. This New Year's morning I was roughly waked by Lavinia shaking me and exclaiming, "Oh, look there!

There's your aunt in bed with Caroline." Seeing two persons asleep in the bed, I jumped out and ran to the right side of it. There lay my aunt, a little on her right side, fast asleep, with her mouth a little open. I recognized her worked nightgown and cap. I stood bewildered, with a childish sort of wonder as to when she could have come; it must have been after I went to bed at night. Lavinia's cries awakened Caroline, who as soon as she could understand, caught the curtains on each side and pulled them together over her. I tore them open, but only Caroline lay there, almost fainting from fright. This lady, Miss Cumberland, afterward became Mrs. Part, wife of a celebrated doctor at Camden Terrace [and now deceased].

"I never talked of what had occurred, but one day after I had returned home, I said to my mother, "Do you know, Mamma, I saw Auntie when I was at school." This led to an explanation, but my mother instead of commenting upon it, went and fetched her mother saying to her, "Listen to what this child says." Young as I was, I saw they were greatly shocked, but they would tell me nothing except that when I was older I should know all. The day came when I learned that my dear aunt suffered dreadfully from the noise of St. Bride's bells, ringing in the New Year. My father tried to get them stopped but could not. Towards morning she became insensible; my mother and grandmother seated on either side of her and holding her hands, she awoke and said to my mother, "Now I shall die happy, Anna, I have seen my dear child." They were her last words.

"(Signed) D. E. W.'"

No general register of deaths was kept at the time of the incident here related, and we have exhausted every means to discover a notice of the death, without success. But we have procured a certificate of Mercy Cox's burial, which took place on January 11, 1829. This is quite compatible with the statement that the death was on January 1st (though such an interval, even in winter, is no doubt unusual), as the lady was buried in a family vault, and probably a lead coffin had to be made. January 1st would be, at the very least, a day of very critical illness. As to the date of the apparition, the marked character of New Year's Day decidedly favors the probability that Miss W.'s memory is correct.

In answer to inquiries, Miss W. says:

"'I was born in 1819. The death of my aunt took place in 1829. Though to my most intimate friends—as Sir Philip Crampton, the late Earl and Countesses (2) of Dunraven—I have often mentioned the event (and to Judge Halliburton), I think I never wrote it fully except for Lord Dunraven and his mother, in 1850, who were very desirous to publish it, but I declined. I think that a great reason I have always had for not talking of it was the awe with which it inspired my mother, and her strict commands that "I should not mention it to anybody." Then, too, I went to school and lost sight of Lavinia Cumberland, and I shrank from the comments of strangers."

"In conversation, Miss W. added that she had never experienced any other hallucination; also that the Cumberland girls had visited her home, and seen her aunt—which accounts for Lavinia's recognition of the figure.

"[We learn through a relative of Miss Lavinia Cumberland that she herself does not recall the incident; but that she remembers hearing her sister speak of a 'ghost case' in which they had both been somehow concerned.]"

The following case Mr. Myers contributes to "Phantasms of the Living" (Vol. II, p. 305), and he remarks that it is a narrative of whose accuracy there is no reason to doubt, as the narrator, Dr. O. B. Ormsby—who wrote from a place called Murphysborough, Illinois, U.S.A., in 1884—had been in communication with Mr. Myers, and replied to his questions. The narrative, which I abridge, is as follows:

"In 1862 Dr. Ormsby was acting as Assistant Surgeon to the 18th Illinois Volunteers; the regiment having gone forward to attack Fort Henry, he was left behind in charge of the sick. Among these was a young man called Albert Adams, a sergeant-major, in whom the doctor seems to have been specially interested. He removed him from the hospital and took him into a private house; the adjoining apartment to that occupied by the patient was divided from his room only by a thin partition; this other room was occupied by the doctor's wife.

"The man was dying and all the afternoon he could only speak in whispers; his father was sent for, and at 11 p.m. Sergeant Adams to all appearance died. Dr. Ormsby, who was at the time standing beside the father by the bed, states that, thinking the bereaved man might faint in the keenness of his grief, he led him away to a chair in the back part of the room, and himself returned to the bedside, intending to close the eyes of Adams, who he thought had expired. Dr. Ormsby then states:

"As I reached the bedside, the supposed dead man looked suddenly up in my face, and said, 'Doctor, what day is it?' I told him the day of the month, and he answered, 'That is the day I died.' His father had sprung to the bedside, and Adams turning his eyes on him said, 'Father, our boys have taken Fort Henry, and Charlie (his brother) isn't hurt. I've seen mother and the children, and they are well.'

"He then gave comprehensive directions regarding his funeral, speaking of the corpse as 'my body,' and occupying, I should think, as much as five minutes. He then turned towards me and again said, 'Doctor, what day is it?' and I answered him as before. He again repeated, 'That's the day I died,' and instantly *was* dead. His tones were quite full and distinct, and so loud as to be readily heard in the adjoining room, and were so heard by Mrs. Ormsby.

"(Signed) O. B. Ormsby, M.D."

In reply to further questions, Dr. Ormsby wrote that he had no opportunity to learn whether what was said about the mother and children was correct, but that he learned afterward that Fort Henry was taken, and the brother was uninjured.

CHAPTER V
MUSIC HEARD AT THE TIME OF DEATH BY THE DYING OR BY PERSONS PRESENT AT A DEATH-BED

Among the numerous cases in which music is heard at the time of death, the following incident, well attested by different observers, is quoted from "Phantasms of the Living," Vol. II, p. 639:

A master of Eton College, Mr. L., wrote to Mr. Gurney in February 1884, enclosing a memorandum made shortly after the death of his mother, which occurred in 1881.

It appears that at the time of her death, several persons were present in the room: the Matron of Mr. L.'s house (Miss H.), a middle-aged, experienced woman; the doctor in attendance (Dr. G.); a friend of the dying lady (Miss I.); and two other persons (Eliza W. and Charlotte C.).

Immediately after Mrs. L.'s death, Miss H. and Charlotte C. left the room to procure something, and shortly after they had left, Miss I. heard a sound of "low, soft music, exceedingly sweet, as of three girls' voices." It seemed to come from the street and passed away. Dr. G. also heard it and went to the window to look out. No one could be seen outside in the street. Eliza W., who was in the room, also heard a sound as of "very low, sweet singing." Mr. L. himself, who sends the memorandum, heard nothing. The two others who had left the room, Miss H. and Charlotte C., distinctly heard the sound of singing as they were coming upstairs.

Later on, when those present were talking over the matter, they found that each one of them had heard the sound of singing and music—*except Mr. L.*

It was specially noticeable that the staircase, up which Miss H. and Charlotte C. were coming, was at the back of the house and away from the street. The time of Mrs. L.'s death was about 2 a.m. on July 28, 1881.

In reply to inquiries, Miss I. sent the following memorandum, which she made *immediately after* the death of her friend, Mrs. L.; it is as follows:

> "*July 28th*, 1881
>
> "Just after dear Mrs. L.'s death, between 2 and 3 a.m., I heard a most sweet and singular strain of singing outside the windows; it died away after passing the house. All in the room [except Mr. L.] heard it, and the medical attendant, who was still with us, went to the window, as I did, and looked out, but there was nobody. It was a bright and beautiful night. It was as if several voices were singing in perfect unison a most sweet melody which died away in the distance. Two persons had gone from the room to fetch something and were coming upstairs *at the back of the house* and heard the singing and stopped, saying, 'What is that singing?' They could not, *naturally*, have heard any sound from outside the windows in the front of the house from where they were at the back.
>
> "E. I."

Dr. G., who was in attendance upon Mrs. L., writes to Mr. Gurney in 1884 as follows:

> "Eton, Windsor
>
> "I remember the circumstance perfectly. I was sent for about midnight, and remained with Mrs. L. until her death about 2:30 a.m. Shortly after, we heard a few bars of lovely music, not unlike that from an aeolian harp—and it filled the air for a few seconds. I went to the window and looked out, thinking there must be someone outside, but could see no one, though it was quite light and clear. Strangely enough, those outside the room heard the same sounds, as they were coming upstairs quite at the other side of the house."

Mr. Gurney adds a note that as Mr. L., although present at the time of his mother's death, did *not* share the experience of the others, this is strong evidence that the sounds did not come from any persons singing outside the house, and the other evidence quoted confirms this.

There are, however, many cases in which the dying persons or those near the bedside have heard musical sounds that could not be attributed to any earthly source. These sounds may have their origin, in some cases at least, in the minds of the living.

The following case appears to point to a hallucinatory origin of the music heard. It is an interesting case and worth quoting in an abbreviated form. It is printed in the "S.P.R. Journal," Vol. IV, p. 181.

Here the subject was a deaf mute, John Britton, who was taken dangerously ill with rheumatic fever, which caused his hands and fingers—which were his only means of conversation—to become so swollen that he could not use them, greatly to the distress of his relatives, to whom he could not make known his wants nor his sufferings.

The narrator, Mr. S. Allen, Steward of Haileybury College and a brother-in-law of John Britton, states that the doctor, thinking John could not recover, had sent for members of his family. He adds that when he and his wife were in a room below John's bedroom, they were greatly surprised to hear music coming from upstairs and ran up at once to find out what it was. He narrates as follows:

> "We found Jack lying on his back with his eyes fixed on the ceiling, and his face lighted up with the brightest of smiles. After a little while Jack awoke and used the words 'Heaven' and 'beautiful' as well as he could by means of his lips and facial expression. As he became more conscious, he also told us in the same manner that his brother Tom and his sister Harriet were coming to see him. This we considered very unlikely as they lived some distance off, but shortly afterwards a cab drove up from which they alighted. They had sent no intimation of their coming, nor had anyone else. After Jack's partial recovery, when he was able to write or converse upon his fingers, he told us that he had been allowed to see into Heaven and to hear most beautiful music."

Mr. Allen asks, "How did John know that Tom and Harriet were traveling, and how could he have heard these musical sounds which we also heard?" He remarks that the music could not have come from next door or from the street, and he gives a rough plan of his house to show that it was not in a row, and that the sounds could not be due to any normal cause.

Mrs. Allen confirms her husband's statement and says that she heard the sounds of singing which came from her brother's bedroom, and that when she entered the bedroom, he was in a comatose state and smiling, and his lips were moving as if he were in conversation with someone, but no sound came from them. Mrs. Allen continues, "when he had recovered sufficiently to use his hands he told me more details of what he had seen, and used the words 'beautiful music.'" She adds that her brother died a few years later, and states "the nurse and I were watching in the room, my brother was looking just as he did on the former occasion, smiling, and he said quite distinctly and articulately 'Angels' and 'Home.'"

The Rev. L. S. Milford, a master at Haileybury College, in giving an account of the interview he had with Mr. and Mrs. Allen, states that "Mrs. Allen says the sounds she heard resembled singing—sweet music without distinguishable words—that she went upstairs directly she heard the music, which continued until she reached the bedroom. Mr. Allen's impression is that the sound resembled the full notes of an organ or of an aeolian harp."

The following interesting case is an instance in which the dying person heard the sound of singing and also had a vision of a lady of whose death she was unaware. The case is taken from the "Proceedings S.P.R." for 1885, [1] and is as follows somewhat abridged:

Mrs. Z., wife of Col. Z. (a well-known Irish gentleman who does not wish his name published), was having some friends to stay with her and asked a Miss X., who was training as a professional singer, to spend a week with her and help to entertain her guests. This she did. Several years later Mrs. Z. became very ill and expected to die; she was, however, perfectly composed and in the full possession of her senses, and was anxious to arrange some business affairs. For this purpose, her husband came to her bedside and talked over these matters with her. Suddenly she changed the subject and said to her husband, "Do you hear those voices singing?" Col. Z., who narrates the incident, replied that he did not, and his wife continued, "I have heard them several times today, and I am sure they are the angels welcoming me to Heaven, but," she added, "it is strange, there is one voice among them I am sure I know, but I cannot remember whose voice it is." Suddenly she stopped and said, pointing straight over her husband's head, "Why, there she is in the corner of the room; it is Julia X. She is coming on; she is leaning over you; she has her hands up; she is praying. Do look; she is going." Her husband turned around but could see nothing. His wife then said, "She has gone."

1 See "Proceedings S.P.R.," Vol. Ill, pp. 92, 93; also "Human Personality," Vol. II, p. 339.

These things the Colonel at the time believed to be merely the phantasms of a dying person, but two days afterwards on taking up "The Times" newspaper, he saw recorded in it the death of Julia, who some years previously had married a Mr. Webley. He was so astounded that a day or two after his wife's funeral he went to see Julia's father and asked if his daughter were really dead. "Yes," he said, "poor thing, she died of puerperal fever, and on the day she died she began singing, and sang on and on till she died."

In a subsequent communication from Colonel Z., the following facts were given:

Mrs. Webley (*née* Julia X.) died on February 2, 1874.

Mrs. Z. (wife of Colonel Z.) died on February 13, 1874.

Colonel Z. saw notice of Mrs. Webley's death on February 14, 1874.

Mrs. Z. never was subject to hallucinations of any sort.

Mr. Gurney subsequently received a note from Mr. Webley (husband of Julia) in which he stated that beautiful as his wife's voice was, it never had been so exquisitely beautiful as when she sang just before her death.

John Bunyan relates an incident of this kind which is worth quoting, though its evidential value is not very great. He states:

> "Talking of the dying of Christians, I will tell you a story of one that died some time since in our town. The man was a godly old Puritan, for so the godly were called in times past. This man, after a long and godly life, fell sick of the sickness whereof he died. And as he lay drawing on, the woman that looked to him thought she heard music, and that the sweetest that ever she heard in her life, which also continued until he gave up the ghost. Now, when his soul departed from him the music seemed to withdraw, and to go further and further off from the house, and so it went until the sound was quite gone out of hearing." [2]

2 See Bunyan's "Works." Edited by George Offor, Vol. Ill, pp. 653, 654. Glasgow, 1855.

CHAPTER VI
VISIONS OF THE SPIRIT OF A DYING PERSON LEAVING THE BODY

The following case, which is taken from the "S.P.R. Journal" (Vol. XIII, pp. 308-311), was sent to Dr. Hodgson by Dr. Burgess, an Associate of the American S.P.R. The vision was seen only by the husband of the dying woman and by none of the others present in the room. The doctor who was present, Dr. Renz, testifies that the percipient, Mr. G., "was in a perfectly normal state before and after, and that there were features in the vision that would not have been likely to occur to him." The percipient, Mr. G., states as follows:

> "My wife died at 11:45 p.m. on Friday, May 23rd, 1902. Gathered around the bedside were some of our most intimate friends, the physician in attendance, and two trained nurses. I was seated at the bedside holding my wife's hand. . . . Earlier in the evening, at 6:45, I happened to look towards the door, when I saw floating through the doorway three separate and distinct clouds in strata. Each cloud appeared to be about four feet in length, from six to eight inches in width, the lower one about two feet from the ground, the others at intervals of about six inches. . . . Slowly these clouds approached the bed until they completely enveloped it. Then, gazing through the mist, I beheld standing at the head of my dying wife a woman's figure about three feet in height, transparent, yet like a sheen of brightest gold; a figure so glorious in its appearance that no words can be used fitly to describe it. She was dressed in the Grecian costume, with long, loose and flowing sleeves—upon her head a brilliant crown. In all its splendor and beauty the figure remained motionless with hands uplifted over my wife, seeming to express a welcome with a quiet, glad countenance, with a dignity of calmness and peace. Two figures in white knelt by my wife's side, apparently leaning towards her; other figures hovered about the bed, more or less distinct.
>
> "Above my wife, and connected with a cord proceeding from her

forehead, over the left eye, there floated in a horizontal position a nude, white figure, apparently her 'astral body.' At times the suspended figure would lie perfectly quiet, at other times it would shrink in size until it was no larger than perhaps eighteen inches, but always the figure was perfect and distinct. . . .

"This vision, or whatever it may be called, I saw continuously during the five hours preceding the death of my wife. All through those five hours I felt a strange feeling of oppression and weight upon my head and limbs. . . .

"At last the fatal moment arrived; with a gasp, the astral figure struggling, my wife ceased to breathe; she apparently was dead; however, a few seconds later she breathed again, twice, and then all was still. With her last breath and last gasp, as the soul left the body, the cord was severed suddenly and the astral figure vanished. The clouds and the spirit forms disappeared instantly, and, strange to say, all the oppression that weighed upon me was gone; I was myself, cool, calm, and deliberate, able to direct, from the moment of death, the disposition of the body, its preparation for a final resting-place.

"I leave my readers to determine whether I was laboring under a mental delusion caused by anxiety, sorrow, and fatigue, or if a glimpse of a spirit world of beauty, happiness, calmness, and peace was granted to my mortal eyes."

The doctor who was present writes as follows:

"From my own observations I can most positively put aside a temporary acute state of hallucinatory insanity during the time of the vision just recorded. . . . I knew Mr. G. well, and I had occasion to know that he never read anything in the occult line; that everything that was not a proven fact was incompatible with his positive mind—so much so that during his vision (of which I did not know at the time) he asked me frequently if I thought he was going to become insane. . . .

"As soon as Mrs. G. was dead, Mr. G., who for six hours was sitting almost motionless next to her, rose and gave all his orders in such a calm and business-like way that it surprised all who were present. If he had labored under a hallucination his mind would not have become clear as suddenly as it did. It is now 2 1/2 weeks since the death and the vision. Mr. G. is absolutely normal physically as well as mentally. He has attended to his business as usual, and, besides, fulfilled many extraordinary duties.

"(Signed) C. Renz"

Many well-authenticated cases are on record where the relatives of a person, watching by the death-bed, have seen at the moment of death a cloudy form rising from the body of the deceased and hovering for a time in the room and then passing away.

Lady Mount Temple informed me that something of this kind was noticed by a psychic friend of hers, who was present at the death of Lord Mount Temple. Others present did not see it.

In a letter that has recently been sent to me of a late well-known dignitary of the Church (a Dean) in New South Wales, he describes the death of his son a few years ago.

He says that at about 3:30 p.m. he and his wife were standing one on each side of the bed and bending over their dying son, when just as his breathing ceased they both saw "something rise as it were from his face like a delicate veil or mist, and slowly pass away." He adds, "We were deeply impressed and remarked, 'How wonderful! Surely that must be the departure of his spirit.' We were not at all distracted so as to be mistaken in what we saw."

The following cases are recorded by Mrs. Joy Snell, in her book, "The Ministry of Angels":

"It was about six months after I began work in the hospital that it was revealed to me that the dying often really do see those who have come from the realms of spirit life to welcome them on their entrance into another state of existence.

"The first time that I received this ocular proof was at the death of Laura Stirman, a sweet girl of seventeen, who was a personal friend of mine. She was a victim of consumption. She suffered no pain, but the weariness that comes from extreme weakness and debility was heavy upon her and she yearned for rest.

"A short time before she expired I became aware that two spirit forms were standing by the bedside, one on either side of it. I did not see them enter the room; they were standing by the bedside when they first became visible to me, but I could see them as distinctly as I could any of the human occupants of the room. I recognized their faces as those of two girls who had been the closest friends of the girl who was dying. They had passed away a year before and were then about her own age.

"Just before they appeared, the dying girl exclaimed, 'It has grown suddenly dark; I cannot see anything!' But she recognized them immediately. A smile, beautiful to see, lit up her face. She stretched forth her hands and in joyous tones exclaimed, 'Oh, you have come to take me away! I am glad, for I am very tired.'

"As she stretched forth her hands, the two angels extended each a hand, one grasping the dying girl's right hand, the other her left hand. Their faces were illumined by a smile more radiantly beautiful even than that of the face of the girl who was so soon to find the rest for which she longed. She did not speak again, but for nearly a minute her hands remained outstretched, grasped by the hands of the angels, and she continued to gaze at them with the glad light in her eyes and the smile on her face.

"Her father, mother, and brother, who had been summoned that they might be present when the end came, began to weep bitterly, for they knew that she was leaving them. From my heart there went up a prayer that they might see what I saw, but they could not.

"The angels seemed to relax their grasp of the girl's hands, which then fell back on the bed. A sigh came from her lips, such as one might give who resigns himself gladly to a much-needed sleep, and in another moment she was what the world calls dead. But that sweet smile with which she had first recognized the angels was still stamped on her features.

"The two angels remained by the bedside during the brief space that elapsed before the spirit form took shape above the body in which physical life had ceased. Then they rose and stood for a few moments one on each side of her, who was now like unto themselves; and three angels went from the room where a short time before there had been only two."

"About a month after the death of Laura Stirman, which I have just related, another friend of mine died in the hospital, a Mr. Campbell, a man of 45. It was pneumonia that carried him off. He was a good and devout man and for him death held no terrors, for he was sure that it was but the transition to a happier, more exalted life than can be lived here. His only regret at dying was that he would leave behind him a dearly-loved wife; but that regret was softened by the assurance that their parting would be only for a time, and that she would join him some day in that other world whither he was going.

"She was sitting by his bed, and, believing as he believed, was awaiting the end with resignation. About an hour before he died he called her by name, and pointing upwards, said, 'Look, L——, there is B——! He is waiting for me. And now he smiles and holds out his hands to me. Can't you see him?' 'No, dear, I cannot see him,' she replied, 'but I know that he is there because you see him.' B—— was their only child who had been taken from them about a year before, when between five and six years of age. I could plainly see the little angel with curly flaxen hair and blue eyes, and garbed in what I call the spirit robe. The face was just that of a winsome child, but etherealized and radiant as no earthly faces ever are.

"The father had been greatly weakened by the ravages of his disease, and the joyful emotion occasioned by seeing his angel child seemed to exhaust what little vitality he had left. He closed his eyes and sank into a placid sleep. He remained in that state for about an hour, the angel child meanwhile staying poised above the bed with an expression of glad expectancy on his radiant face. Occasionally he looked lovingly at his mother.

"The breathing of the dying man grew fainter and fainter until it ceased altogether. Then again I witnessed what had now become a familiar spectacle to me—the formation of the spirit body above the discarded earthly body. When it was complete the angel child grasped the hand of the now angel father, each gazed into the eyes of the other with an expression of the tenderest affection, and with faces aglow with joy and happiness they vanished.

"Later on in the day, the widow (Mrs. Campbell) said to me, 'I am very glad my dear husband saw B—— before he died; it was natural that B—— should come for him to take him to the angels, for they loved each other dearly. I shall now be able to think of them as always together and happy. And when I receive my summons I know that they will both come for me.'

"After I had left the hospital and had taken up private nursing I was engaged to nurse an old lady (Mrs. Barton, aged 60), who was suffering from a painful internal disease. She was a widow and her only daughter lived with her... The time came when the end was very near. The mother had been for some time unconscious, and the daughter was kneeling by the bedside, weeping, her face buried in her hands. Suddenly two angels became visible to me, standing on either side of the bed. The face of one was that of a man who, when he departed from this life, was apparently about 60 years of age. His beard and hair were iron-grey; but there was stamped on his features that indescribable something indicative of exuberant vitality and vigor, which shines forth from all angel faces I have seen, whether in other respects they present the semblance of youth or old age. The face of the other angel was that of a woman, apparently some ten or fifteen years younger.

"The dying woman opened her eyes, and into them there came that look of glad recognition I have so often observed in those whose spirits are about to be released forever from their earthly tenements. She stretched forth her two hands. One angel grasped one hand and the other angel the other hand, while their radiant faces were aglow with the joy of welcoming to the better world her whose earthly pilgrimage was finished.

"'Oh, Willie,' she exclaimed, 'you have come to take me home at last, and I am glad, for my sufferings have been hard to bear and I am very tired.' Then she added, 'And you too, Martha!' With the joyous light still in her eyes her hands remained outstretched for perhaps half a minute. Then they seemed to slip from the grasp of the angels. All her sufferings were over.

"The daughter had raised her head at the sound of her mother's voice, and her tear-dimmed eyes seemed to reflect something of the glad surprise depicted on her mother's face.

"'I can doubt no more after this,' she said to me when her mother had breathed her last breath; 'I know that mother saw father and her sister, Aunt Martha. I know that they came to take her to her rest in heaven.'

"Eagerly she listened to me when I told her a little later how I had seen two angels depart with her angel mother. 'I believe it! I believe it!' she cried, 'but, oh, how I wish that I could have seen it too!'"

Creative Thought and the Problem of Evil

By

Prof. Sir W. F. Barrett, F.R.S.

SECOND EDITION

REVISED AND ENLARGED

CONTENTS

PREFACE

THIS little book formed the substance of an address delivered to the Quest Society at Kensington Town Hall some four years ago. It was published in the *Quest Quarterly Review* for July 1910, and afterwards issued in a small volume. The first edition printed being exhausted, I have revised and somewhat enlarged the essay, and it is now issued in a more permanent form. Obviously, a brief essay such as this does not pretend to touch more than the fringe of a large and important subject, which needs fuller exposition. But I hope that what is said may suggest lines of thought perhaps new and helpful to inquirers.

When this lecture was delivered, I had not seen Professor H. Bergson's great work on *l'Évolution créatrice,* an English translation of which has recently been published. Bergson, whilst rejecting the mechanistic view of evolution, does not accept that of final causes—that is, of a conscious end to be attained. He holds that the unity of life is to be found solely in the primal impulse that pushes it along the road of time, the harmony not being in front but behind; that Nature is not a plan in course of realization, for a plan closes the future whose form it indicates, whereas in evolution the portals of the future remain wide open. Life, he argues, is ever evolving before us in a continuous creation of unforeseeable forms, a creation that goes on forever in virtue of an initial movement. "At a certain moment," Bergson says, "in certain points of space, a visible current has taken rise; this current of life, traversing the bodies it has organized, one after another, passing from generation to generation, has become divided amongst species and distributed amongst individuals, without losing anything of its force, rather intensifying in proportion to its advance." This ever operative and transcendent impetus of life, meeting first with the resistance of inert matter, overcame it by degrees, and formed it to be receptacles and organs of life.

With great force and wealth of knowledge, Bergson demonstrates how all mechanical theories of evolution fail hopelessly. He uses the wonderful structure of the lenticular eye as an illustration, supporting the argument I touched upon in my address (see p. 46). Doubtless, adaptation to the environment is a condition of life and helps explain some of life's twists, but it does not explain its onward movement, any more than a riverbed explains the moving river. "It is one thing," as Bergson says, "to recognize that outer circumstances are forces evolution has to reckon with, another to claim that they are the directing causes of evolution. The latter theory is that of mechanism"; but while he shows that the evolution of life is other than a series of adaptations to accidental circumstances, he holds, as already remarked, it is *not* the realization of a plan.

How, then, does Bergson account for the mystery of, for example, the organ of vision? He points out that the eye, while it is an organ of infinite complexity of structure, is one of extreme simplicity of function. He holds that the simple act of vision had to be attained, and this act has organized, not manufactured, a vast assemblage of elements to produce vision of the objects upon which the being can act. Bergson compares "the process by which Nature constructs an eye to the simple act by which we raise our hand." If the hand be raised through a resisting medium, the greater the effort, the further it will advance; but wherever it stops, the medium will so far shape itself to the motion. Thus he conceives the marvelous complexity of the organ of vision to be the material impress of the simple progress towards vision. The form of the organ only "expresses the degree in which the exercise of the function has been obtained" ; if the progress towards vision has gone equally far in widely different animal species, the form of the visual organ will be similar.

But although Bergson denies that there was any predetermined plan in the life movement, he admits that its progress "implies at least the rudiments of choice," and therefore "the anticipatory idea of several possible actions." In fine, he remarks that "the evolution of life looks as if a broad current of consciousness had penetrated matter, loaded as all consciousness is, with an enormous multiplicity of interwoven potentialities." And again he says, with regard to the formation of an organ of vision, "Whether we will or no, we must appeal to some inner directing principle to account for this convergence of effects." But if this primal life impulse has consciousness, the power of choice, and an inner directing principle, it differs little from the characteristics of what I have called in this essay the transcendent Creative Thought, ever immanent and operative within the organism.

Bergson, however, will not admit that any predetermined purpose or plan exists in Nature. He maintains there is no Supreme Reality behind phenomena except that which is found in Mobility, in Becoming. "All reality," he says, "is tendency, if we agree to mean by tendency an incipient change of direction." But if a Supreme Mind be not the Ultimate Reality, as I have striven to maintain in this essay, it seems impossible to account for the fact that the "tendency" of evolution is ever towards higher and more complex forms of life, or to account for the intricate structure of, *e.g.,* the eye, if there is no Supreme Mind in which pre-exists the plan whereby vision is to be attained. The argument I have ventured to put forward in Chap. IV. on behalf of a Mind above Nature appears to me more consistent with our experience and common sense than the elusive and aimless continuity of Becoming which arises from Bergson's life impulse.

At the same time, we must admit that evolution is not only a progressive and upward movement: in some cases, it ceases to progress, and in some cases seems to deviate or turn back, as if blindly groping its way. But, as Bergson himself points out, we must beware of extending to the things of life the same reasoning and modes of explanation which we find so sure and successful in inert, unorganized matter. The way in which life goes to work to attain a certain result, we constantly find, is just the way which we should never have guessed beforehand; all *a priori* reasoning here fails.

Thus also the difficulties and contradictions we meet with in life, referred to in Chap. V., may largely arise from our trying to make the varied phenomena of life, with its consciousness and volition, fit in with the forms of thought, the theory of knowledge, and the reasoning, which we have derived from, and apply with success to, inert matter. We seek in life, as Bergson remarks, the same kind of symmetry and order that we find in the physical world, the world of matter and space, the world of geometry and logical sequence. But the true theory of life transcends all these concepts of the physical world and of logical reasoning, and will unfold itself only when we have attained to a far deeper consciousness, a more profound intuition, than is possible for us to possess amid the limitations of our present existence.

CREATIVE THOUGHT
CHAPTER I

THE SEEN THE OFFSPRING OF THE UNSEEN. THEORIES OF EVOLUTION. TELEPATHY

EVERYONE recognizes that the physical basis upon which materialism has so long rested has been undermined, if not destroyed, by recent discoveries in the very branch of science upon which that narrow school of thought had built its confident assertions. For generations it has been the accepted axiom of science that the elementary sub- stances of which the material world is composed were imperishable, immutable, eternal realities. To-day this axiom cannot be relied upon. For some of the elements, and others will follow, have already proved to be slowly undergoing transformation from more complex to simpler states ; and the dreams of the alchemist may well revive. Even the atoms themselves are no longer, as Lucretius and his successors thought, "strong in their solid singleness," but have vanished into a congeries of electrons. Matter, which appears so solid and stable, has been proved to be nothing else than hidden *molecular motion*, and every phenomenon of the external world is reducible to molecular motions of one kind or another. Hence the crude materialistic view of the universe, with all its arrogant assumptions, instead of being a popular, has become a damaged and discredited theory.

Endless evolutionary processes in the physical world, beginning no one knows how, and blindly going on no one knows whither, cannot explain the visible universe. For behind—and the source of all tangible matter—lies the unseen, intangible, incomprehensible *Ether;* and behind—and the source of all physical energy—lies an unseen, all-pervasive and incomprehensible *Force.*[1] The progress of science is in fact steadily pushing back the boundaries of the seen, and compelling us to believe, as we were told long ago, "that what

1 See note 1 at end, pg. 100..

is seen hath not been made out of things which do appear,"[2] but is the offspring of an unseen universe, and an unseen, indwelling, and transcendent Power.[3]

On the other hand, the visible universe is not, as theologians used to teach, a wonderful contrivance created as we now see it, and then set going by the fiat of the Almighty. Even in the controversies of the recent past, our orthodox religious teachers might have recognized that creation by *evolution* is a far more amazing fact than creation by *fabrication*. A man can make a steam- engine or a motor-car, but he cannot make them slowly evolve themselves out of iron ore and other natural products, nor repair themselves when they get out of order, nor reproduce them- selves when they decay. But in organic Nature, from the simplest living cell to the most complex living organism, we find these three processes ever at work—a *formative* power, a *reparative* power, and a *reproductive* power. Moreover, the evolutionary processes, whilst preserving and perpetuating life, are ever pressing onward to higher types and higher faculties of life. What was once *explicitly* declared to be the Divine purpose in the spiritual world," that they may have life and may have it abundantly,"[4] is, in the natural world, the principle which biologists tell us *implicitly* underlies all evolution.[5]

Doubtless some biologists, with Weismann, say that this principle is purely mechanical and that life is nothing more than "a chemico-physical phenomenon" ; but to create and keep going a series of "living reproductive machines," "with ever-widening faculties, they have to invent inscrutable factors, "determinants" or "dominants" contained in the germ-plasm, "by whose fluctuations

2 Heb. xi. 3 (R.V.)

3 For an interesting and suggestive book on the spiritual interpretation of the Cosmos see Mr Stanley Redgrove's Matter, Spirit and the Cosmos, published by Rider & Sons, London. 2s. 6d.

4 St John x. 10.

5 Since the first edition of this essay was published, my eminent friend the veteran naturalist, the late Dr A. Russel Wallace, O.M., F.R.S., etc., issued his remarkable work *The World of Life*. In this deeply interesting volume Dr Wallace gives us the conclusions to which his long and profound study of Nature has led him. He unhesitatingly affirms—and illustrates with a wonderful wealth of knowledge—that beneath and beyond all evolutionary processes lies an unseen Directive Power, or hierarchy of intelligent powers, in the unseen. It is gratifying to find that, in the main argument of my essay, which he kindly read in proof, I am in agreement with one so much more competent than myself to form an opinion on this profound biological problem.

in a plus or minus direction the appropriate variation is attainable."[6] Others wisely recognize their ignorance of the mysterious directive power behind evolution; some assume "the existence of original, innate capabilities of the living substance, not dependent on selection," and others frankly assert that there exist "unknown factors in evolution." But whatever be the unseen, unknown, and immanent Power behind, it presents all the characteristics of purposive guidance, having a definite aim, and therefore I prefer to call this inscrutable factor "Thought."

May I therefore venture to ask you to consider with me the subject of the "Creative and directive power of Thought," not, however, as a philosophical speculation, but as a scientific conclusion, based upon experimental evidence? There is no need to illustrate this statement as regards our own conscious thought. For the works of the savant, the musician, the architect, the painter, the preacher, and the poet are monuments of this creative power. And is it not true of all of us, that we project our ideas, our thoughts, into the unseen, and they come back to us as realities in the seen? To a large extent we create our own world. The man who achieves has *faith* in something above and beyond himself. It is faith that "subdues kingdoms, works righteousness, and obtains promises." Doubt, distrust, pessimism, are tokens of the lack of faith, and the result is an arid and sterile life, aimlessly criticizing all things and accomplishing nothing.

But the dynamic of thought extends beyond well-recognized actions. There is abundant experimental evidence that one mind can act upon another mind independently of the recognized channels of sense. This was first noticed by the early mesmerists and called by them "community of sensation" ; although ridiculed by the physiologists and psychologists of the day, the fact is now undeniable. It is over forty years ago since my attention was drawn to the existence of this faculty, by some hypnotic experiments which I witnessed when staying at a country house in Ireland. The experiments were made by a friend on a subject who proved to be exceedingly sensitive to hypnosis, and who subsequently was hypnotized by myself. Though skeptical at first, I was convinced, after carefully repeating the experiments myself,

---

6 Weismann, *The Evolution Theory*, ii. 346. See a most able and lucid reply to the Darwin-Weismann theory of life by Mr T. W. Rolleston in his admirable work *Parallel Paths*, published by Duckworth & Co., which I warmly commend to any who have not read it.

that anything I felt, or tasted, or saw, was (under strict test conditions) usually perceived by the subject. Repeating the experiments in my own house, later on, the fact was confirmed. Hence it was that in 1876 I read a paper before the British Association detailing these and other facts and urging the appointment of a committee to investigate the phenomenon of what I ventured provisionally to term "Thought-transference," and which to-day is now widely accepted and known as "Telepathy."[7]

It was the not unnatural disinclination of the recognized scientific societies to publish the record of these early investigations of mine on "Thought-transference," that suggested the need of a Society for "Psychical Research," in the foundation of which in 1881-2 a few friends warmly co- operated with me. Evidence soon accumulated that not only could one mind evoke a more or less faithful image of a definite concept or percept existing in another mind without the aid of any of the recognized channels of sensation, but that certain persons could also project and create a phantasm of themselves perceived by a distant friend! That is to say, the distant agent A could not only evoke a consciousness of himself or of his thought in the percipient B, but also occasionally enable B to perceive the presence of A, though the latter was some miles away. In the quite independent experiments which my friends the late Rev. W. S. Moses, M.A., and Mr S. H. Beard conducted, under the strictest conditions to avoid collusion, conclusive evidence was obtained that an act of thought in the mind of A could create an apparently real external image of A in the mind of B. These facts corroborated similar occurrences well known to students of Eastern mysticism, and they furnish an obvious explanation of many—though I do not think of all—of the well-established cases of apparitions at the moment of death; the telepathic impact, by direct *psychosis,* creating the apparition of the distant person.

There is really nothing so very extraordinary or amazing about this. All that we perceive in the world around us is the projection of a series of phantasms from our own minds, created, it is true, by some stimulus received through one or other of the organs of sense. Take the case of vision. External objects form a minute inverted image of themselves on the retina, as they do on a photographic plate in the

7 A brief summary of some of the evidence on behalf of Telepathy will be found in my little volume on Psychical Research published in the "Home University Series," Williams & Norgate, price 1s.

camera; but we do not *look* at the retinal image as we look at the ground-glass screen in a camera. The retinal impression is transmitted to the brain, a certain tract of nerve cells is thereupon stimulated; this multitude of separate stimuli our *ego* collects into a coherent whole, and we then mentally project, *outside ourselves,* a phantasm of the object, apparently as big and erect as the object really is. But that appearance is the creative act of our own mind and leads us, rightly or wrongly, to think that the appearance is a real objective thing resembling the thing in itself. It is a good working hypothesis and practically useful, but it is a mental projection for all that. The thing *in itself*, or the world of Divine thought, may be, and doubtless is, wholly unlike the percept and sensory phantom we habitually create.

The fact is, we invariably "confound the organs of perception with the being that perceives. The eye is not that which sees, it is only the organ by which we see; the ear is not that which hears, but the organ by which we hear; and so of the rest."[8] This being so, it is not very surprising to find that a telepathic impact from a distant mind may lead the recipient to project an apparently external image of the distant person; or to find, as in the well-known cases of post-hypnotic suggestion, that a suggestion made to a subject under hypnosis, *e.g.* that he will see So-and-so at a certain time, causes the hallucination to be evoked so vividly that it has all the appearances of a real objective presence. In like manner a discarnate mind may be able telepathically to impress those living on earth so that apparitions of the dead may be evoked, or other sensory hallucinations produced. This also seems to me, in our present state of knowledge, the most tenable theory of many genuine cases of "hauntings."[9]

8 Reid's *Inquiry into the Human Mind*, p. 246 (Hamilton's edition). There is a singular coincidence between this passage from Reid and the following from Swedenborg's *Divine Love and Wisdom*, § 363, published the year before, viz. 1763. "It is according to the appearance that the eye sees. On the contrary, it is the intellect that sees by means of the eye. It is an appearance that the ear hears, but it is the intellect (*mens*) that hears by means of the ear. . . . And so of the rest." Indeed it is as true now as when Reid wrote that "no man is able to explain how we perceive external objects any more than how we are conscious of those that are internal," unless indeed we accept the theory of "spiritual influx," as Malebranche and Swedenborg maintained.

9 A discussion of the various theories of hauntings is given in Chapter XIII of my little book on *Psychical Research* already referred to.

CHAPTER II

THE TENDENCY OF THOUGHT TO EXTERNALIZE ITSELF. THOUGHT AND LANGUAGE ; SPIRIT AND MATTER

PHANTASMS, however, are *mental* creations, and the question arises, Can thought produce a change in organic structure? As a matter of fact it does so every time we will to move our muscles; the physiological processes thus involved have been explored, but the mystery of the interaction of mind on matter, the interplay of soul and body, remains as insoluble as ever. Moreover, thought can set going creative and reparative processes in the physical organism. Of this there is abundant evidence. But before adducing some facts in support of this statement, let us ask ourselves what we mean by the word *Thought.* Obviously it embraces far more than definite ideas clothed in speech or action. In fact, does not consciousness, whether in man or animals, carry with it the conception of perceptive power, feeling, and volition, and thus imply the existence of thought? Doubtless sustained rational thought is the prerogative of man, but we cannot withhold thought and even some reasoning power from animals. Romanes has shown this to be true of domestic animals, and we could all cite instances indicative of thought and reason in our domestic pets. It would be interesting to obtain evidence as to whether our thoughts or emotions can be transferred, say, to a favorite dog, telepathically. That animals sometimes see phantasms and are in abject terror from this cause appears to be undeniable.[10] But can we excite a telepathic response in the higher animals ? Will a dog answer to its name if silently called? Is there any evidence of community of emotion and sensation between master and dog, apart from the ordinary channels of sense perception? If so, we might get into closer touch with the animal world, and perhaps in the future this may be attained, for all life is one in its hidden Being.[11]

Many naturalists have given numerous instances of what appear to be the beginnings of reason and thought in the animal creation;

10 For evidence of this see *Proceedings* of the Society for Psychical Research, vol. x. pp. 326-330.

11 Is this what is foreshadowed in Rom. viii. 19, that the "earnest expectation of the creation waiteth [with longing and expectancy] for the revealing of the sons of God"? Or is it the hidden Sonship of humanity awaiting its full emergence?

and even in plant-life there seem to be indications of intention and directed purpose. Moreover, what in our ignorance we call *instinct* in animals is surely evidence of inarticulate *thought* that may, and often does, transcend our own intelligence. The migration of birds and fishes, the homing of certain birds and animals, are familiar instances. Certain creatures and some races of primitive men obviously possess some extraordinary sense which is not recognized as yet by science ; a sense of direction or a transcendental perceptive power would seem to exist, though it may not awaken any act of consciousness. Professor Watson, of Johns Hopkins University, has recently experimented with certain tropical sea-birds—a species of tern which make their nesting-place, and most northern habitat, in the Gulf of Mexico. Half a dozen of them he took from their nests, marked each with different-colored paint, and their corresponding nests likewise. He then sent them by a friend, in cages in the hold of a steamer, to be liberated near New York. They were set free at the assigned spot on a given date, and five days later most of them found their way back to their nests, after traversing upwards of 1000 miles over sea and coast they had never before visited. To call this "instinct" is merely to cloak our ignorance. It is certainly to us a *super-normal* perceptive power, which awakens in the creature a form of thought, impelling to action; but probably not through any reasoned or conscious process.

That a directive force can be given to our own muscles without the intervention of consciousness, physiologists know perfectly well and call it "reflex action" as in the acts of walking, breathing, the beating of the heart, etc. But what I wish here to assert is that above and beyond ordinary and known reflex action, a *transcendental perceptive power,* which stimulates unconscious and yet purposive action does really exist. It is difficult in the case of human beings to obtain evidence of this, owing to the intrusion of ideas or sensory impressions. But in certain persons, especially those of limited education and ideas, we can and do obtain evidence of the possession of this unconscious, intelligent, and transcendental perceptive power. The motion of the so-called divining—or dowsing-rod is merely the outward and visible sign of this inward and singular gift. Any object hidden in an unknown place, whether underground water, mineral lodes, or buried coins, a good "dowser" may unconsciously discern and locate, by a sudden muscular spasm, which twists his forked rod with a vigor he cannot control nor consciously reproduce. It would take me too far from my present subject to give the abundant evidence upon which this assertion rests. It will be found in the reports on the so-called divining-rod which I have published in the *Proceedings* of the Society for Psychical Research.

Hence the term Thought, whilst usually restricted to the intellect, has really a much wider significance than its mere expression in words and acts. It includes not only the ideas, feeling and will, that prompt conscious and voluntary acts, but also the subliminal stimulus that creates involuntary, unconscious, and yet purposive acts. Of absolute or pure thought we know nothing whatever, nor can we form the smallest conception of what it is, except as an aspect of the Universal within the creature.[12] All we know of thought is its *expression* within the limitations of our present existence. Behind the expression lies the wordless idea or emotion, which far transcends the clothing of the idea in words or acts. But this we do find, *thought ever seeks to externalize itself*, and it is this manifestation we perceive. May we not therefore say that thought is an inscrutable, directive power, seeking to express and thus to realize itself?

In fine, the characteristic of thought, so far as we know it, is that which creates some purposive action, whether consciously in the world without, or unconsciously to us in the world within ourselves. As regards our conscious life, Language is the expression of articulate thought; and Art—pictorial, plastic, or musical—the expression of inarticulate thought, or *emotion.*[13] Language and Art are therefore the modes, the instruments, whereby human thought manifests itself and becomes known to itself. Even when the mental processes are wordless, they would have been impossible unless the mind had previously been trained to the use of muscular or verbal signs.[14]

Hence we cannot separate a conscious thought or emotion from its *expression* in some form of muscular action or language, or its depiction in some form of art. Thought and language used in its widest sense, whilst essentially distinct—and in the relation of prior and

12 Philosophers in all ages have attempted variously to define and describe thought; it is needless to give quotations. As one truly says: "By the faculty of thought we approach still nearer to the supreme intuitions of the soul."

13 See Note 2 at end, pg 101.

14 It is true that certain mental processes, termed "intuitional thinking," seem to be independent of language, but they have their basis in some form of mental imagery, as in playing chess or the use of algebraic symbols.

posterior, of cause and effect—are however to us indissoluble, and cannot be conceived of apart.

In like manner matter is a mode in which Spirit expresses itself, and matter is as dependent on Spirit for its existence as language is dependent on thought. The two are to us unthinkable apart: we cannot conceive of matter without Spirit, nor of Spirit without the attributes of matter. The apparent contradiction and antinomy of matter and Spirit disappear when we recognize that in language and thought we have the same apparent antithesis—though we all realize there is no real antagonism here, but that language is the utterance and self-realization of thought. And just as the great poet or artist knows that his composition is only an inadequate expression of the thought he strives to realize, so the material world is, to us, only a partial realization of Spirit, one aspect of the Universal Thought.

And if the laws of Nature, if the wonderful diversity and mystery of life, if infinite electrons and molecules, suns and stars and the great galaxy of heaven, be the channels cut in our present material world by the Creative Thought of God, what limitless wonders, what new forms of life and intelligence, may we not expect to find, when we can apprehend the operation of the Divine impulse on the more plastic material of an unseen universe?

CHAPTER III

THE POWER OF SUGGESTION. HYPNOSIS. THE SUBLIMINAL SELF

LET us now inquire what evidence exists of the dynamic action of thought in creating definite physiological and structural changes in our own organism. The therapeutic value of suggestion, now so widely recognized by the medical profession, is an illustration of this. Anyone who has read the pages of the *Zoist,* or the more modern and masterly treatise of Dr Lloyd Tuckey on " Suggestive Therapeutics," will find cases of the cure not only of nervous and functional disorders, but cases of tissue-change, of organic processes, produced by suggestion.[15]

15 *Treatment by Hypnotism and Suggestion*, by C. Lloyd Tuckey, M.D. (Baillière & Co., 1907, 5th ed.). See, e.g., pp. 74, 76, 78, 389 ff.

In hypnosis the patient is more readily under the influence of suggestion, inasmuch as any conscious resistance to the idea implanted is in abeyance, and accordingly more profound and successful therapeutic effects can thus be produced than in the normal waking state. Conscious *assistance* to the idea suggested can be aroused by expectation through, *e.g.*, conversation with a patient cured of a like disorder by hypnotic treatment, or by religious conviction as in faith-healing, or at a sacred shrine such as Lourdes. Equally efficient are any religio-philosophical views, if we think them inspired, as in Christian Science. Though I gladly recognize the help and healing which this creed has afforded to many, yet it may be said of Christian Science that what is new in it is not true, and what is true in it is not new. For precisely similar cures were wrought at certain pagan shrines in ancient Rome, where votive offerings to Esculapius were dedicated to the god for the miraculous cures that had been wrought.[16]

History is full of the miracles of healing accomplished by suggestion. All will remember the cures by Greatrakes and by Prince Hohenlohe in Germany; of these latter, a century ago, a contemporary and eyewitness wrote: "The deaf were made to hear, the blind to see, and the lame to walk, not by any medical art, but by a few short prayers and the invocation of the name of Jesus," etc. We say these miracles are due to *suggestion*, but that again is only a word to conceal our ignorance of the processes involved. They are the result of a conscious thought or suggestion in one person, liberating the subconscious recuperative and formative forces within the organism of the patient. That there *is* a hidden self below the threshold of consciousness appears to be abundantly proved. This has been termed the *subliminal self;* but whether we accept this hypothesis or not, there can be no doubt that there are high transcendent powers within us far more wonderful and more subtle than any conscious acts of thought.

Take, for example, the processes of nutrition. Can anything be more mysterious than the conversion of our food into muscle, bone, and nerve? A sleepless directive and selective force is ever at work within ourselves. Ask the most learned chemist with all his appliances to extract milk from a bundle of hay; he would laugh at the idea; but give the hay to the humble cow and this mysterious selective force within herself works this miracle. *How* we know not—only that the

16 See the excellent article on "Mind Cure" etc., by Dr and Mr F. W. H Myers, in Proc. S.P.R., vol. ix. pp. 160 ff.

infinitely minute molecules of hay must have been pulled to pieces and rearranged into the molecules of milk. Talk about miracles; can anything be more amazing and improbable than this? And yet we pass by this familiar fact as if the handling and building up of molecules (millions of which lie in the smallest microscopic speck) were as simple and easy a thing as the handling of bricks to build a house. I repeat, some transcendent directive and selective power, beneath and beyond the level of consciousness, apparently exists within us each, and as this power is purposive and intelligent it has all the characteristics of thought. This being so, we shall not be surprised to find that profound physiological processes can be induced by this subliminal power—this self within ourself.

It may seem a trivial thing to adduce as an illustration of this the removal of warts by suggestion, but nothing is trivial in the arcana of the Unconscious in bodily life. Nearly fifty years ago the great physiologist Dr W. B. Carpenter stated that "the charming away of warts by spells belongs to cases which are real facts however they may be explained." Three hundred years before this, Lord Bacon mentions, in his *Natural History,* a cure thus produced upon himself. He tells us that he had from childhood a large and unsightly wart upon one of his fingers which no physician could cure. When sixteen years old he went to Paris and (I quote an abbreviation of his own words) "there grew in a month's space on both my hands at least a hundred warts. The English ambassador's lady, a woman far from superstitious, told me she could remove my warts if I would do what she directed. Thereupon she got a piece of lard, touched the warts with it, including the large one I had from childhood, put the remainder of the lard in the sun, on a window-sill which faced south, and assured me that as the lard melted away all the warts would also disappear. At this I did greatly marvel, but it was as she said: all the warts went quite away, and even that large one I had so long endured." Bacon attributed the result to the doctrine of sympathy, then the fashionable hypothesis among the learned to explain any mystery.

Dr Hack Tuke, in his classical work, the *Influence of the Mind upon the Body*, gives other illustrations of a similar kind. Thus, the daughter of a surgeon whom he knew had about a dozen warts on her hands; caustic and other remedies had been tried without success for eighteen months, when a friend told her, after solemnly counting the warts, that they would all disappear the following Sunday, which actually came to pass. I might quote many instances of a similar kind attested by careful observers and physicians of eminence. Miss Mason, who was till lately the senior lady inspector of boarded-out children, a personal friend of mine, cured a little boy, terribly afflicted with warts, by a simple charm.[17]

Hence it is unquestionable that *suggestion,* taking the form of faith , or words, or charms, can sometimes do what no medical skill can accomplish. That terrible diseases such as hydrophobia can be both caused and cured by suggestion, has been established beyond any doubt.[18]

I myself have seen rapid and startling physiological changes, impossible to produce by any conscious act, brought about by a mere suggestion made to a patient under hypnosis. Thus a red scar or a painful burn, or even, *mirabile dictu*, a figure of a definite shape, such as a cross or an initial, names can be caused to appear on the body of the entranced subject solely through suggesting the idea. By creating some local disturbance of the bloodvessels in the skin, the unconscious within us has done what it would be impossible for the unassisted conscious self to perform. And so also in the well-attested cases of *stigmata*, where a close resemblance to the wounds on the body of the crucified Savior appears on the body of the ecstatic. This is a case of *self*-suggestion, arising from the intent and adoring gaze of the ecstatic upon the bleeding figure on the crucifix. A state of incipient hypnosis, or mono-ideism, is thus produced, the voluntary powers are inhibited, and with the abeyance of the conscious self the

---

17 This and other cases are fully detailed in the *Journal* of the S.P.R., vol. viii. See also Miss Feilding's excellent book on *Faith Healing* (Duckworth & Co.) .

18 See Dr Hack Tuke, *Influence of the Mind upon the Body*, pp. 202-216. As these pages were passing through the press I received evidence of some remarkable cures recently performed in Russia, by "spiritual healing," through the Rev. A. M. Niblock; even cripples, it is stated, were cured, as by Prince Hohenlohe and also a few years ago at Knock in Ireland. In like manner the Rev. S. J. Whitmee, whom I know personally, relates in a little book now before me, the numerous amazing cures of natives in Samoa, which he effected twenty years ago by the use of Count Mattei's remedies, "blue electricity," etc., minute harmless globules, in which no drug whatever can be detected by analysis.

The thaumaturgy of a simple and sincere faith is only now being recognised, and is not yet adequately used as a therapeutic agent by the medical profession. A new significance is now given to the words: "All things are possible to him that believeth." "Whatsoever ye pray and ask for, *believe that ye have received them,* and ye shall have them."

hidden powers emerge, whilst the trance and mimicry of the wounds are strictly parallel to the experimental cases above referred to.

Again, in the lower forms of life, change of environment sometimes appears to have a suggestive effect in producing color- changes. For the beautiful experiments of Professor Poulton, F.R.S., have shown that some caterpillars can more than once in their lifetime change their color to suit their surroundings. Thus, if one half of a set of certain green caterpillars have black twigs placed among the leaves on which they feed, and if the other half have some white paper spills placed among the leaves, most of the former will become black and most of the latter white. The nervous stimulus which produces these different pigmentary deposits appears to be excited by the particular color acting upon the surface of the skin. But through what wonder-working power is the change brought about? Not, of course, through any conscious action of the caterpillar, for the pupae of these same caterpillars undergo a like change, changing even to a golden appearance, with a brilliant metallic lustre, when the chrysalis has been allowed to remain on gilt paper! Nothing, of course, occurs if the chrysalis be dead; but when living its dormant life, the unconscious within is apparently able to respond to suggestion from without; the color of the environment having this suggestive effect.

May not a similar cause be at work in the many cases of protective mimicry, as well as protective coloration found in the animal kingdom? If we accept the usual biological explanation of protective mimicry, the long intermediate stages required by natural selection would render the creature in many cases not less but *more* conspicuous among its kind, and therefore expose it to greater danger of capture and less chance of survival. In fact, I am convinced that biologists have too long closed their eyes to the *psychic factor in evolution*—the directive power of the unconscious within the organ- ism. Evolutionary processes in Nature are, according to this view, the expression of the creative power of thought, using the term in the wider sense already defined. But it is *thought immanent, operative and transcendent, within the organism.* And it is interesting here to recall the fact that one hundred and fifty years ago Swedenborg—who was a true seer as well as a learned man of science—explicitly urged this very hypothesis of an inherent directive force in the development of the forms of life.[19] A century later E. von Hartmann, in his well-known work the

19 Thus in his *Economy of the Animal Kingdom,* § 275, he writes: "We must ak-

Philosophy of the Unconscious, developed much the same view, only he rejects all anthropomorphic ideas, or any form of consciousness or personality in the Supreme apart from Nature, whereas Swedenborg's theology is the reverse of this. Von Hartmann with great wealth of learning shows that in the phenomena of instinct and of clairvoyance (which latter he states, as did Schopenhauer, only the ignorant reject) we have additional evidence of the Unconscious in life.

Is it therefore illegitimate to assume that the Unconscious, the psychic factor in evolution, is operative in the process of development to bring about more perfect organs and higher types of life? It seems to me impossible to explain, *e.g.*, the development of the mammalian eye according to the usual method adopted by biologists . It would take me too far aside from my present purpose to trace out this argument ; I can only state, after a prolonged study of the subject of vision, that to attempt to explain the production of the lenticular eye by the action of blind and fortuitous evolutionary forces, or endless ages of natural selection , appears a more incredible hypothesis than the assumption of an unconscious or rather a Superconscious directive Thought ever operative within the organism.[20]

As Professor Calderwood has said: "There is a Power operating continually in nature which does not come within the range of the observation possible to scientific modes and appliances, yet to which science is ever indirectly bearing witness."[21]

---

nowedge, if we think of causes and origins, that such a directive or formative force is not without but within the chick or embryo; and that it must exist within that substance that was first in the ovum and that has life or soul within it," etc. In fine he tells us, "the Infinite is in the finite, as in receptacles." Moreover, now that telepathy may be regarded as a *vera causa,* every living cell in the organism (as Mr Gerald Balfour has conceived) is possibly in telepathic *rapport* with every other cell, and our unitary consciousness may be the resultant of this *rapport* among the brain cells. The wide philosophical implications of telepathy have not yet been adequately discussed.

20 The bifocal eye of the Brazilian fish anableps is inexplicable upon a purely naturalistic theory. Mr T. W. Rolleston, in his suggestive work Parallel Paths, has referred to the anableps' eye, pp. 100 ff., and shown how impossible it is for natural selection, laying hold of fortuitous variations, to explain this remarkable organ. My friend the late Mr W. F. Kirby, F.L.S., has pointed out to me that some water beetles have also bifocal lenses in their eyes, the upper lenses looking above the water in which they swim and the lower lenses below the surface of the water.

21 *Man's Place in Nature*, p. 341. See also Professor Bergson's view referred to in the Preface, p. 9, and Dr Driesch's Gifford Lectures, or his later work *The Problem of Individuality*, p. 19, etc., where he maintains that the vital processes depend on a "supra-personal" non-mechanical agency.

CHAPTER IV

THE DIVINE MIND

ARE we, then, driven to the conclusion that there is no Divine Mind outside of Nature, but that Absolute Thought is only *feeling its way*, as it were, to self-realisation? Let us try to grapple with this difficulty. Philosophy teaches us that the phenomena of Nature are only signs and symbols significant of something we cannot fully apprehend with our limited senses. We can watch and explore the wonders of Nature, just as we can watch the movements of a telegraphic needle and learn to read the message it brings, but the moving needle does not enable us to perceive the operator at the other end who is causing it to move, nor does it even remotely resemble the operator. The telegraphic signals give us, it is true, an intelligible message, but *it is intelligible only because the intelligence of the operator has been and is related to our intelligence.* In like manner the mental signs our brain and nervous mechanism give us of the material world outside are not the things, nor a resemblance to the things in themselves; the real world around us, the world of *ontology,* the world of *Divine Thought,* is absolutely inaccessible to us. But the reason why the material world is intelligible, why we can interpret the signs it gives us, is because there is an Intelligence behind the universe which has been and is related to our intelligence. In other words, an Infinite Mind exists and is related to our mind. For the interpretation of Nature reveals an orderly and *intelligible* system, and we are never put to intellectual confusion in our scientific researches, as would be the case if the universe were a chaos and not a cosmos. Hence in the universe, external to ourselves and to all finite forms of life, there exist Intelligence, Will, and Purpose. The Ultimate Reality must be Mind, and therefore God is Transcendent Mind. This is the conclusion reached by great thinkers in all ages.[22] Nature "is not a soulless interaction of atoms, nor life a paltry misery closed in the grave."

---

22 In recent times we might quote to the same effect Professor Challis, Sir John Herschel, Professor Clerk Maxwell, Sir G. Stokes, Lord Kelvin, and other eminent physicists in our own country, as well as philosophers like Lord Haldane (see his Gifford lectures on Reality).

The Creative Thought of God is thus unfolding and realizing itself in the worlds . and lives around us. This is the Eternal Word, the Logos, through whom the worlds were made, and in whom "we live and move and have our being." And if this be so, then, to quote Swedenborg once more, "the Deity is in each single thing, and this even to such an extent that there is in it a representation of the Eternal and Infinite. . . . From this influx arises effort, from effort force, and from force the effect."[23]

But if Nature be the unfolding and indwelling of the inscrutable Creative Thought, why is there so much in Nature that appears to us imperfect, objectionable, and evil? As regards noxious creatures no doubt they fulfill their function in the universal arcana of life, as does every microbe, whether, from the human point of view, it be malignant or benevolent. It is our restricted vision that is chiefly at fault, and gives rise to our mistaken reasoning (see the last paragraph in the Preface).

Moreover, we must remember that the varied aspects of life and thought are integral parts of an *organic whole.* A *single* aspect of life and thought may be quite unintelligible and even appear mischievous and contradictory to another aspect, when viewed apart from the whole. Is it not so with our own thoughts and actions? How purposeless, unintelligible, and foolish, the isolated action of the best and wisest of men might appear to a casual onlooker with restricted knowledge! We are ever apt to forget the limitations of our intelligence and that we can never grasp, nor conceive, the *totality* of life and thought. Even in our own personality our conscious thought is but a fragment of our whole personality and only an item in the great

---

23 *Arcana Cœlestia,* § 5116. In like manner Emerson says in his essay on "Self-Reliance": "We lie in the lap of immense Intelligence, which makes us receivers of its truth and organs of its activity. When we discern justice, or truth, we do nothing of ourselves, but allow a passage to its beams. If we ask whence this comes, if we seek to pry into the soul that causes, all philosophy is at fault. Every man discriminates between the voluntary acts of his mind and his involuntary perceptions, and knows that to his involuntary perceptions a perfect faith is due. This is the ultimate fact which we so quickly reach on this, as on every topic, the resolution of all into the Ever-blessed One." And so also Browning:

"There is an inmost centre in us all
Where truth abides in fulness, and to know
Rather consists in opening out away
Whence the imprisoned splendour may escape,
Than in effecting entry for a light
Supposed to be without."

universe of being.

Our self-consciousness is no doubt the last and highest rung, but it is only one rung, in the ladder of evolution, and the long and toilsome ascent to it embraces infinitely more than is contained in our restricted self-conscious life. Each rung of the ladder in its turn was doubtless a state of semi-consciousness, a region of effort and attainment; and when habitude had rendered that sphere of work effortless and perfect, it passed into unconscious automatic action, thus enabling the ascending impulse of life to take another step upward to still higher aims.[24] The pinnacle of self-conscious life was reached in us when ideas and reason were evolved, when consciousness regarded not only the *me* within but the *not me* in the world around and recognized their distinction. Albeit in subject and object, as in thought and language and in spirit and matter, and in the problems of life, we have a conflict and antithesis, which find their solution in a higher Unity, the manifold expression of the Universal Creative Thought.

CHAPTER V

THE PROBLEM OF EVIL

FINALLY we are confronted with the mystery of good and evil, the sorrow and suffering of human life. This age-long problem has been grappled with by abler men than myself, and it is too profound and difficult a subject adequately to discuss at the close of an address. But a few thoughts present themselves:

(i) Everything in the phenomenal world is subject to limitation, imperfect expression, and apparent contradiction, but we believe all is progressing towards some higher unity as yet unimagined by us.

(ii) Consciousness, whether of physical or moral states, is only aroused through differences, the contrast of opposites. Without relativity of sensation there would be no sense-perception. A uniform state of heat, an unchanging illumination, a monotony of sound,

24 Diverging, as Bergson would say, into the marvellous instincts of insects and animals on the one hand and into the intelligence and reason of man on the other.

would not arouse the sensation of warmth, or vision, or hearing. And so a uniform state of goodness would not enable us to perceive the good. *If there were no wrong to strive with there would be no active right.*

(iii) All our knowledge, we must remember, is *relative,* and the apparent truth or error, the apparent rightness or wrongness, of an action depends upon its harmony with the content of our experience. Even what the noblest deem most true, most right, and most real is therefore, in this sense, only relatively true.

(iv) Evil may be regarded as a *privation or defect in being,* an imperfect, incomplete state of existence. The privation not in itself being wrong, nay being the necessary expression of our creaturely existence, but the *love* of that defect or imperfection being evil, as an offense against God, the perfect Being.[25] Swedenborg called this defective selfhood of man his *proprium,* having no life or truth in itself , and from it originates the falsities of self-love and self-sufficiency. But such defect in us is essential to our freedom of will, our choice of right or wrong, of loving the permanent on the one hand, or the illusory on the other. As all life comes from God, the influx of His Spirit alone enables us to know our defects and pursue the right.

(v) All manifestation, whether of force, or life, or goodness, involves the *overcoming* of some reluctance or resistance. The physical energies of gravitation, electricity, light, heat, etc., reveal themselves to us in the process of overcoming the inertia of matter. We know absolutely nothing of the abstract or essential nature of Force, or of any physical Energy, such as gravity, electricity, etc. We do not see electricity in a lightning flash, nor in the electric light. What we see is the overcoming of resistance in the path of some hidden power; we see the glowing of particles of air in the lightning, or of carbon in the electric lamp—a resultant jointly due to the resistance encountered and the transit of the unseen power. Reduce that resistance, as in the copper conductor, and the light and heat disappear; abolish it altogether, both in gross matter and in the ether, and electricity would be utterly unknown and undiscoverable. Inertia, impedance, resistance, call it what we will, is as essential to the manifestation of

25 The Neoplatonic view that evil has no independent nature was adopted by St Augustine, whose argument that evil is the *privation of good* is well known.

every physical force as the force itself. The most powerful locomotive ever made, the largest cruiser ever built, would be unable to move a fraction of an inch if the *resistance* of the rails or the water were not present. And is not this physical fact exactly similar to the moral problem that is before us? As the forces of Nature reveal themselves to us, and become self-realised, through the stresses and strains they generate in some reluctant medium—so the moral and spiritual forces of the unseen reveal themselves to us, and become realized in us, by the stresses and strains they create in the reluctant or opposing medium of our inert or evil lives.[26]

Here, as elsewhere in the natural and spiritual worlds, "the invisible things of Him since the creation of the world are clearly seen, being perceived through the things that are made."[27] With our present limitations what we term evil seems therefore as real and as necessary a thing for our spiritual life as good; for in overcoming the former the Divine life becomes manifest and our higher self is, to that extent, realized. It is in the *attaining* and not in the attainment, in the *effort* and not in the rest, in the *struggle*, and not in the victory, that the process of self-realisation, the deeper consciousness, the larger and higher life, which we desire and mean by immortality, truly consists.

Intervals of rest, hereafter as well as here, will doubtless alternate with periods of activity. But the continued survival of consciousness implies *desire,* and appears to rest upon *effort*, and therefore doubtless some form of nutriment will ever be required, by the spiritual as by the natural body, to sustain its expenditure of energy.

In every aspect of our present life strength and progress come through continued effort. But effort involves both a hidden and sustaining power and an encountered resistance. We cannot tell what may be the nature of those resistances, corresponding to difficulty and evil in this life, which exist in the unseen world, but if there be any correspondence and continuity, as we all believe there is, between the seen and the unseen, analogous and perhaps more stupendous obstacles may confront the soul in its progress through the ages still to come. Only thus, in ceaseless overcoming, does the Inscrutable and Infinite Thought, the transcendent One behind the manifold, unfold

---

26 The life and work of Florence Nightingale (as of many others) afford an illustration of great achievements won by the mental and moral power she possessed of overcoming the resistances she encountered.

27 Rom. i. 20.

and enshrine itself in higher and more abundant life, of which we see the beginning here on earth but await its fuller revelation through the unending ages in the unseen life beyond.

In conclusion we must, however, remember there is an internal quest, the quest of the soul, as well as the external quest of fuller, larger life and consciousness. This inner quest is the knowledge of God, the consciousness of His presence. Such knowledge is incommunicable from without, and cannot be attained by effort from within. It is, in fact, the voice of the Spirit to the spirit, or, as Philo said, of the "Alone to the alone," for the soul "must be very still to hear God speak." Of this Divine indwelling the humblest human souls have known, this mystic way they have traversed, alike with the greatest prophets and poets.[28]

> "Calm be the body for her in that hour, and the tumult of the flesh.
>
> "Yea, all that is about her calm; calm be the earth, the sea, the air, and let Heaven itself be still.
>
> "Then let her feel how into that silent heaven the Great Soul floweth in.
>
> "And so may man's soul be sure of vision, when suddenly she is sure of light: for this light is from Him and is He; and then surely shall one know His presence when like a god of old time He entereth into the house of one that calleth Him, and maketh it full of light.
>
> "And how may this thing be for us? Let all else go."
>
> *Plotinus* (translated by F. W. H. Myers).

CHAPTER VI

SUMMARY OF THE ARGUMENT

(i) THE inscrutable thing we call our conscious thought is known only in its creative utterance, or partial self-realization, through some form of expression or language, using this term in its widest sense ; that is,

28 See note 3, p. 101.

its manifestation is dependent on something foreign to itself. That which we know *subjectively* as consciousness is *objectively* a conflict between will or desire and some opposing resistance or non-satisfaction.[29]

(ii) There is also a subconscious or subliminal life, which reveals itself in automatic and involuntary processes and muscular movements. This subliminal life has many powers that transcend the self-conscious life, inasmuch as it can control and direct physiological processes and molecular forces to a desired end. In all living things, ourselves included, we see a transcendent directive power at work, and thus we find ourselves linked by a common bond to the whole realm of life and to the Universal life.

(iii) This subliminal activity can also be directed and set going by verbal, telepathic, or emotional suggestion coming from within or without the organism, and thus profound functional and organic changes can be brought about. Suggestion in lower organisms may also come from the environment, and may result in corresponding colour-changes in the creature; thus, in part, may come about protective coloration and mimicry. The psychic factor in evolutionary processes can no longer be neglected, though its importance has hitherto been strangely overlooked. That an unknown directive force, a psychic factor, is present and potent in the development of organs and species, and in the whole course of evolution, appears unquestionable.

(iv) The Super- onscious Creative Thought is thus manifesting itself in the phenomenal world, and through evolutionary processes which are ever tending towards higher and more abundant life.

(v) But in this manifestation, whether animate or inanimate, something—equally inscrutable to us as the Dynamic Thought itself—has to be overcome. In the process of overcoming (which involves effort and a corresponding expenditure of energy), the hidden Thought, the Logos, unfolds itself, and with a fuller manifestation as ages of attaining pass

---

29 E. von Hartmann, in his *Philosophy of the Unconscious* (Eng. ed.), ii. 88, has developed the same view, which, however, was unknown to me when this address was delivered. He writes: "Only when the radiating will meets with a resistance by which it is checked or broken can it lead to objective manifestation of *existence,* to the subjective phenomenon of *consciousness.*" So also Jacob Böhme says: "Nothing can become manifest to itself without repugnance; for it has nothing to oppose it, it is ever going out of itself, and does not return to itself."

by, until consciousness and ultimately self-consciousness are reached in human personality. Then the ascent in moral and spiritual life begins, the attain- ment of one height setting free aim and effort to win another, and thus to develop a deeper fuller content in life.

(vi) The mystery of evil cannot be solved, nor the effects removed, by empty denials of its existence. We are conscious of the limitations and relativity of our knowledge, and believe an ultimate reconcilement of the conflict of good and evil will be found in a far-off transcendent Unity. We see in physical phenomena an analogy to this spiritual conflict, *e.g.*, as the existence of opposite polarities is inherent in certain forces such as magnetism, or as the contrast of opposites such as light and shade is needful to arouse perception, so our moral consciousness is awakened by the opposing forces of good and evil. Moreover, to confer freedom of choice between good and evil, a *privation* of perfect being is necessary to our selfhood; it is the loving of this privation, this illusion, instead of God, the source of all being, that is evil, not the existence of the privation. Further, as all manifestations of physical energy are due to the operation of an inscrutable force overcoming an inscrutable resistance, so spiritual life and energy are quickened within us by our overcoming, through the Divine influx, that inscrutable thing which now presents itself to us as evil.

Hence higher ideals and new difficulties to overcome are likely to be ever before those who would win and retain "the life which is life indeed."

"The energy of life may be
Kept on after the grave, but not begun;
And he who flagg'd not in the earthly strife—
From strength to strength advancing—only he,
His soul well knit and all his battles won,
Mounts, and that hardly, to eternal life. "

Matthew Arnold, whose words these are, was possibly mistake in sayin in saying "*all* his battles won"; they may be, in fact, but just begun.

Nevertheless, in the highest quest of the soul, the quest of the consciousness of God, abandonment of effort, stillness and loss of self, as the mystics have ever taught, is the window of Divine illumination.

"For more than once when I
Sat all alone, revolving in myself
The word that is the symbol of myself,
The mortal limit of the Self was loosed,
And past into the Nameless, as a cloud
Melts into Heaven. . . .
. . . .And thro' loss of Self
The gain of such large life as match'd with ours
Were Sun to spark—unshadowable in words,
Themselves but shadows of a shadow world."[30]

NOTES

1. The Ether And Force, p. 78.

The *Ether* of space is in a sense material, for it belongs to the physical universe, though it is of another order than that of the ordinary ponderable matter with which we are familiar. The Ether has no discontinuity and pervades all things, therefore cannot be perceived by us any more than a deep-sea fish could perceive the water in which it swims. The existence of this incomprehensible Ether, it is true, is, so far, only a creative act of human thought, a useful and necessary hypothesis in the co-ordination of physical phenomena. *Force* is also as mysterious as the Ether. The usual definition of force, given in Newton's second law of motion, is really a vicious circle: for, briefly, that definition asserts that force is the cause of motion, or change of motion, in a right line, *i.e.* of motion of a mass, or momentum, but momentum, *m.v.*, implies the very term "force." Force is really an aspect of stress, and involves the reaction, or resistance to alteration, of the body to which it is applied, and comes into existence with the yielding of that body. Now, the resistance to alteration, or the *inertia* of a body, also involves force as much as does the applied force: inertia only comes into existence with the application of force to a body. Hence force does not exist without resistance, nor resistance exist without force. A moving or a stationary body *un-interfered with* has neither force nor *vis inertia.* Force therefore is as fundamental and as incomprehensible a thing as the Ether itself.

30 Tennyson, *The Ancient Sage.*

2. Emotion And Instinct, pp. 85.

A distinguished school of thought regard emotions as the *effect*, not the cause, of muscular and physiological changes in the organism. That is to say, in all organisms reflex action is produced by certain stimuli , and these reflex activities produce in the higher forms of life characteristic, though imperceptible, sensations, which in turn evoke what we call emotions, each reflex, or combination of reflexes, creating its own particular class of emotion. Hence it is argued the emotion, say of fear, is not that which causes us to tremble or turn pale, but we are afraid because a certain stimulus has caused the reflex action of trembling or turning pale. In like manner Instinct is regarded as the habit of selecting certain of these reflex activities for definite ends, which have been found useful. In course of time these impulses due to instinctive or reflex acts generate experience, and thus, as Lloyd Morgan says, instinct "is organic behavior suffused with awareness." But whilst both emotion and instinct, physiologically considered, are always associated with reflex activities, nevertheless *psychologically* considered they are unquestionably far more than the mere growth of particular reflex activities.

3. The Divine Indwelling, p. 97.

This last paragraph is quoted from an earlier book of mine, and did not appear in the previous edition of *Creative Thought*. The omission was pointed out to me by a valued friend, who writes to me as follows:—

"I have come to see that an attitude of mind of quiet silent reception, apart from reading and study, tended more towards inner enlightenment. Before this I had been too eager, too restless, too prone to think I could teach and develop myself . Since then I have been as quiescent as a naturally excitable person can be. The only effort which I now try to make—if it can be really called an effort—is to keep still and listen, so that now and again in some wonderful moment I may hear the still small voice , which can only be heard in silence.

"Is there not a twofold aspect of spiritual progress? —on the one hand, the strenuous fight and overcoming ; on the other, the silent self - surrender of the soul. The Mystics in all ages, and some at the present day, have taught and realized in their own experience that there is a Divine growth of the soul, quite apart from any effort to attain. To such it has come like the unconscious growth of the body or of the lilies of the field . Rabindranath

Tagore expresses this in *Gitanjali* :—

"On many an idle day have I grieved on lost time. But it is never lost, my lord. Thou hast taken every moment of my life in thine own hands.

"Hidden in the heart of things thou art nourishing seeds to sprout, buds to blossom, and ripening flowers to fruitfulness.

"I was tired and sleeping on my idle bed and imagined all work had ceased. In the morning I woke up and found my garden full with wonders of flowers.'"[31]

"This silent unconscious growth, I believe, takes place even in those who to all outward appearance seem quite depraved. May we not therefore hope that, however worthless and sin-stricken anyone's life here may seem to be, could we but know *all* their present and past experiences, both conscious and unconscious, we should find in some directions that their soul has grown. And this seems to hold the promise that the struggle against evil, which sometimes appears unending, will in reality be shortened, for its continuance does not alone rest on the conscious efforts of humanity towards spiritual progress.

"Perhaps you know the story of a man who went to stay with a very strenuous, hardworking Rector of a parish. After a continuous round of services, meetings, lectures, Sunday and week-day Bible classes and conferences, the visitor said to the Rector, 'This is all very energetic and noble on your part, but you do not seem to leave anything for the Holy Spirit to do in your parish.'"

31 *Gitanjali*, § 81, Macmillan & Co.

SWEDENBORG:
THE SAVANT & THE SEER

BY

SIR W. F. BARRETT, F.R.S.

LATE PROFESSOR OF EXPERIMENTAL PHYSICS IN THE ROYAL COLLEGE OF SCIENCE FOR IRELAND

EMANUEL SWEEDENBORG

PREFACE

In publishing the accompanying lecture by request, a few words of explanation are necessary. The Swedenborg Society—through my old and esteemed friend Mr. J. Howard Spalding—honored me, though not a member of their body, with an invitation to deliver a public lecture on the great Swedish seer. This was given in London during the spring of the present year, and the substance of my address was published in the *Contemporary Review* for July 1912. I have to thank the editors of that review for permission to reprint the article, to which I have made some additions. My cordial thanks are also due to the Swedenborg Society for the admirable arrangements they made for the lecture, to His Excellency Count von Wrangel, the Swedish Ambassador, who presided on the occasion, and to the large and representative audience who honored me by their presence.

It is impossible to have even a slight acquaintance with Swedenborg, or with the land of his birth, without a profound admiration for the man and a deep affection for his country. To no foreign land is an Englishman drawn by so many ties of affinity. But I confess I felt ashamed of my insular ignorance in mixing with the educated classes of Sweden, who converse with freedom in many European languages, and whose graceful hospitality seems an integral part of the charm of their country. Moreover, one is amazed to find how, in spite of its comparative poverty and sparse population, Sweden has set an example to all nations of what an enlightened government can do for the welfare of its people. The educational system of Sweden is practically free and so admirable that it throws our own completely into the shade. To the generous hospitality and valuable guidance of my kind host in Stockholm, Herr Ernst Beckman, I desire to express my warmest thanks.

I am indebted to the Swedenborg Society for the portrait which forms the frontispiece of this little book.

W. F. BARRETT.

KINGSTOWN, CO. DUBLIN.
Oct. 1912.

Note. — Some readers may not be interested in the scientific work of Swedenborg, and may therefore prefer to pass from p. 108 to p. 115.

SWEDENBORG: THE SAVANT AND THE SEER

Sweden held a commanding position in the civilized world when Emanuel Swedenborg was born in January 1688. During the forty years' rule of Gustavus Vasa in the sixteenth century, Sweden first awoke to a national life. Under his grandson, Gustavus Adolphus—the hero of the Thirty Years' War and one of the most enlightened monarchs that ever wore a crown—Sweden rose to be one of the greatest powers in Europe, a position she held for nearly a century. Moreover, Sweden was the champion of intellectual freedom, civil and religious liberty, as shown by the fact that such learned heretics as Descartes, Grotius, and others were welcomed at the Court of Queen Christina. Nor can science forget the debt she owes to those famous Swedes, Rudbeck, Linnæus, Celsius, Berzelius, and others in the past, and in recent years to Angström, Retzius, Nordenskiöld, Sven Hedin, Arrhenius, and many others.

In the great University of Upsala, founded in 1477, Swedenborg's father, Jesper Swedberg, was Professor of Theology and afterwards a Bishop of the Lutheran church, so that an atmosphere of learning and piety surrounded the seer from his youth. Upon completing his University career at Upsala in 1710, Swedenborg travelled widely, spending much time in England, "studying Newton daily," as he writes to his brother-in-law. When he returned to Upsala in 1715, he devoted himself to the study of science, especially metallurgy, and the following year was appointed Assessor at the Board of Swedish Mines by Charles XII. Two years later he was raised to the House of Nobles for his distinguished contributions to science and engineering, and thereupon his family name was changed to Swedenborg. His boundless mental activity and industry continued throughout his long life; for he lived to the age of eighty-four, when he died in London in the year 1772.

Swedenborg's life is divided into two great epochs. The first two-thirds were devoted to the pursuit of science and the investigation of almost every aspect of the natural world; the last third of twenty-eight years, after what he describes as "the opening of his spiritual sight," to the records of that Divine illumination and intromission to the spiritual world which he affirms were vouchsafed to him.

Swedenborg As A Scientific Man

From his earliest years, Swedenborg was an eager and profound student. He diligently strove to master every branch of knowledge, not merely by book work, but by practical investigation. In this, as in other subjects, he was, as the great Swedish chemist Berzelius remarked, "far in advance of his age." It was as rare then, as now, to find a man combining the gift of classical scholarship, and of abstruse philosophical speculation, with an eminently practical and inventive genius. Yet this was the case with Swedenborg. Not only did he make himself acquainted with various skilled handicrafts, but he projected several novel and useful inventions, including a flying machine, a submarine war vessel, a quick-firing gun, a mercurial air pump, and other inventions of recent times. His mechanical genius is better remembered by his successful transportation of war galleys for seventeen miles overland, during the attack on the fortress of Frederickshall. He threw himself with ardor into mathematics, issued a mathematical journal, published in ten books the first Swedish treatise on algebra, and was offered, but declined, the chair of mathematics in his University. Then he began to study nature. Professor Nordenskiöld, in an address to the Swedish Academy of Sciences, tells us that the merit of having first made the fluctuation of the ocean level the subject of scientific investigation belongs to Swedenborg. Geology and paleontology were sciences that did not then exist; rocks, minerals, and fossils were believed to have been created *in situ,* or accounted for by the Noachian deluge. Swedenborg, though at first accepting the current view, soon realized its absurdity, and urged that mineral veins were probably deposited from solutions which penetrated the fissures of the rocks, and that fossil flora were ancient plants and ferns embedded in a clay matrix: in fact, a genus of fossil plants has been named after him "Swedenborgia."[1]

1 The Swedish Academy of Sciences are now publishing a series of Swedenborg's

Swedenborg next turned his attention to cosmology. To understand his writings, which, as usual at that period, were published in Latin, we must remember the great influence which Descartes then exercised on earnest students of nature, as well as on philosophic thought. History repeats itself. Just as a generation ago the works of Darwin and the evolutionary hypothesis were denounced by the orthodox clergy on all sides, so two centuries earlier the Cartesian controversy shook the religious and philosophical world to its foundations.

Nowhere was the effect felt so keenly as in the University of Upsala immediately prior to the birth of Swedenborg. The House of Priests took up the question in the Swedish parliament and tried, but in vain, to exclude all Descartes' disciples from chairs in the University. They urged that the investigation of astronomy and the physical sciences generally should be subject to the dicta of the Church and not allowed to endanger the faith of students by conclusions and theories derived from purely human observation and experiment. In vain they tried to appoint the orthodox members of the theological faculty at Upsala as censors of the whole University curriculum. The only result of such sacerdotal pretension was, as is always the case, to create greater freedom of thought; followed, as Mr. Stroh points out, by "a direct stimulus to unfettered philosophical and scientific research." In fact, this ecclesiastical controversy mainly led to the foundation, in 1710, of the "Royal Society of Sciences" in Upsala, among the early members of which was Emanuel Swedenborg.

It is therefore not surprising to find Swedenborg's scientific writings so largely penetrated with the Cartesian philosophy. Space, according to Descartes, was not a vacuum, but filled with endless vortices of colliding particles of matter, whereby the finer particles were formed by attrition, and from these again still finer, and so on, until all space became a plenum of material particles and cosmic dust. These particles possessed no essential property beyond extension, being all of the same kind. The properties we perceive in different elementary substances Descartes held were due to varieties of size and motion. Motion in this plenum being possible in the same way as the

chief scientific papers, with introductions by eminent savants. English readers are under obligation to Mr A. H. Stroh for editing and translating these Introductions, to which Mr Stroh has added a learned paper on the "Source of Swedenborg's Early Philosophy of Nature."

motion of a fish in an ocean of water. The other school of thought, represented by Newton, held that space was a vacuum through which were scattered impenetrable particles of different kinds of matter, like the stars in the heavens, all matter being united by the force of gravitation. But how this force could act across a pure vacuum Newton could not adequately explain, nor has any solution yet been given. It was mainly this difficulty of "action at a distance" through a void, that led Continental philosophers of that period to prefer the Cartesian view of a plenum, and in this view Swedenborg shared.

Descartes' philosophy, however, led to dualism—to an unbridgable gulf between mind and matter, between Nature and Spirit, between the Finite and the Infinite. Swedenborg saw this, as Leibnitz did fifty years earlier. Leibnitz derived all matter from infinitely minute points or *monads,* each of which mirrored a phase of the universe, of the mind of God, between whom and the sum of created *monads* which made up nature, Leibnitz held there existed a pre-established harmony.

In some respects, Swedenborg's conception of the physical universe resembles that of Leibnitz, whose writings he had studied, though he did not adopt the theory of pre-established harmony. But Swedenborg's view of the origin of matter, though overlaid with complexities and erroneous ideas, is more like that to which science is tending at the present day. He conceived all matter as ultimately derived from what he terms "natural points"—which are, as it were, intermediate between the finite and the infinite. From the varied aggregation and motion of these points and their derivatives, he believed the physical universe was built up according to mechanical and geometrical laws. In his *Principia,* he points out how the rapid motion of a minute corpuscle can generate a line, and the line an area, and the area a solid, and he goes on to say, "a corpuscle thus moving can represent by its celerity and direction something which previously had no existence, and which is quite different from the corpuscle itself: and it is every way a figure, so far as our senses are concerned, although it is merely motion which produces the effect, or by means of motion form is fixed."

Moreover, Swedenborg appears to have been the first to suggest the idea of a *vortex atom*; and a vortex theory of matter, in an incompressible and continuous frictionless fluid, has been revived by Helmholtz and Kelvin in our own day. In the physical point, Swedenborg, like Leibnitz, asserts that all finite things are latent; the macrocosm is hidden in the microcosm. In fact, some years later in his *Arcana Coelestia,* he says: "The Deity is in each single thing, and this even to such an extent that there is in it a representation of the Eternal and Infinite. From this influx arises effort, from effort force, and from force the effect." This may come to be the orthodox view of science—for nature is the unfolding and indwelling of the inscrutable creative thought of God.

It will be noted that Swedenborg attributes *effort* to these ultimate entities, the power of, or tendency to, motion, yet not an actual motion in time and space, but, as he remarks, an "internal state—an effort towards motion." Our modern doctrine of potential energy, as distinguished from kinetic energy (*i.e.,* actual motion in time and space), is here implied, and certainly it is a very early conception of that idea. Swedenborg thus considers the origin of matter to be infinitely minute *centers of force*, and he regards these points as the connecting links between mind and matter, between the physical world and the self-existent Inscrutable Source of all things.

Some thirty years after this, an Italian philosopher, Boscovich, published a theory of matter that, in some respects, resembles that of Swedenborg. According to Boscovich, matter is simply an aggregation of mathematical points that are centers of potential energy. Fifty years ago, the greatest experimental philosopher that ever lived, Michael Faraday, was led to much the same view. "Matter," Faraday says, "must fill all space, or at least all space to which gravitation extends, for gravitation is a property of matter dependent on a certain force, and it is this force which constitutes matter."

In his own day, Faraday's view of centers and lines of force, extending through and filling all space, received scant support; but recent scientific research strongly confirms them. The once universal belief in eternal, immutable atoms, scattered in various states of aggregation through empty space, has been replaced by congeries of infinitely minute, swiftly moving *electrons*, which appear again reducible to physical points or centers of electric force, everywhere present, and which by their rapid motion may give rise to the fundamental properties of matter.

Another modern idea held by Swedenborg was that the structure of the physical universe is built upon a common plan, which runs through all things from the least to the greatest. He conceived that the disposition of planets, suns, and stars had their analogue in the arrangement of the molecules of matter. Like Descartes and Faraday, Swedenborg was impressed by the wonderful lines of force around a magnet—revealed in beautiful curves when iron filings are scattered on a sheet of card placed above a magnet. These lines of force are crowded along, run parallel to, the axis of the magnet, and unite each magnetic pole in closed curves. Turning his gaze to the heavens, Swedenborg noticed the innumerable hosts of stars were packed more closely along the Galaxy, or Milky Way, which he therefore conceived to be the axis of the visible universe. He suggested there might be still greater systems in nature of which the Milky Way is itself but an element.

"It is certainly," as Professor Arrhenius has remarked, "a grand thought to attempt an explanation of the world according to which a complete harmony exists between the greatest and the least—the stellar systems and the atoms of matter—or even, according to Swedenborg's conception, its least part, the physical point." Swedenborg's *Principia,* in which these views are expressed, despite its many defects and fallacious a *priori* reasoning, was a work of genius and one of the great achievements of his scientific career. This is remarkably seen in Swedenborg's foreshadowing the *nebular hypothesis,* twenty-one years before Kant and sixty-two years before Laplace. It would take too long to enter into detail on this question, but it is summed up by Arrhenius, who states in his Essay on Swedenborg's cosmology, that:

> "The ideas which were first given expression to by Swedenborg, were afterwards, in a much modified form, consciously or unconsciously, taken up by other authors in cosmology. Swedenborg held that :—
>
> "(1) The planets of our solar system originate from the solar matter ; taken up by Buffon, Kant, Laplace, and others.
>
> "(2) The earth and the other planets have gradually removed themselves from the sun, and then received a gradually lengthened time of revolution ; a view expressed in our own day by Sir G. H. Darwin.
>
> "(3) The earth's time of rotation-that is to say, the length of the day-has been gradually increased ; also the view of Sir G. H. Darwin.

"(4) The suns are arranged around the Milky Way; taken up by Wright, Kant, and others.

"(5) There are still greater systems of which the Milky Way forms a part; taken up by Lambert."

Not only in cosmology but also in physics, Swedenborg anticipated some of our present knowledge. In his day, heat was considered a subtle fluid, caloric; however, Swedenborg regarded it as a mode of motion and also advocated the undulatory theory of light. Regarding magnetism, while many of his theories now appear puerile, his views on the molecular structure of a magnet were singularly correct and ahead of his time. He pointed out that no increase in weight is produced by magnetization, but "the smallest particles of iron are drilled into a straight line; turned round and brought into a definite order." However, he noted that this state is disturbed or destroyed by heat, bending, or blows, which "unsettle the arrangements of the minute parts."

We must not, however, be blind to the fact that Swedenborg's theories of the modes of motion and subdivision of matter, and the terms he employs, "Finites," "Actives," and "Elementaries," now only excite a smile in the scientific reader. Like many other savants of his day, Swedenborg's scientific speculations and reasoning were vitiated by the then prevalent system of deductive philosophy. The ancient philosophers argued that the circle is the most perfect of figures, and hence the heavenly bodies being perfect must move in circular orbits. Swedenborg, in like manner, failed to grasp the supreme significance of Kepler's first law and assigns circular and spiral motions to his material entities because they also are "absolutely perfect." But in spite of much that we may dismiss in the light of modern knowledge, there runs throughout the whole of Swedenborg's *Principia* not only the true scientific spirit of an earnest seeker after truth but a conception of the constitution of matter and of the structure of the universe, which may be regarded, a recent scientific writer remarks, "as a first daring venture into the unknown and as the pointing out of a new road which is now being travelled in chemistry and physics with increasing surety that the goal is in sight."

From cosmology, physics, chemistry, and metallurgy, Swedenborg passed to the study of *biology* and published two great works on the

Animal Kingdom and the *Economy of the Animal Kingdom.* A sympathetic discussion of their value will be found in the *Transactions of the International Swedenborg Congress,* held in July 1910, and published by the Swedenborg Society. Here will be found a series of learned papers by eminent physiologists and anatomists, who point out the surprising anticipations of modern science found in Swedenborg's anatomical works. Thus Professor Ramström, the distinguished Professor of Anatomy in the University of Upsala, deals with Swedenborg's work on the *brain* and shows that Swedenborg was the first to arrive at the conclusion that the grey matter of the cerebrum was the seat of psychical activity, and that through it sensory impressions were transformed into thoughts and volition. Dr. Henschen, the Professor of Pathology in the Medical Faculty of the University of Stockholm, has a learned paper on the localization of functions in the brain and shows that Swedenborg correctly, and even with detail, localized the motor-centres in the brain one hundred and twenty years before this discovery was published by German and English physiologists.

In fact, as the eminent savant, Professor Gustaf Retzius, stated in his Presidential address to the Congress of Anatomists in Heidelberg nine years ago: "Swedenborg was not only a learned anatomist and sharp-sighted observer, but also in many respects an unprejudiced, acute, and deep anatomical thinker." This is evident in Swedenborg's discussion of the ductless glands of the human body, such as the spleen, the thyroid, the pituitary glands, etc. Dr. Goyder, the consulting physician to the Bradford Infirmary, states that "Swedenborg anticipated many of the pre-eminent offices of these ductless glands, which the medical profession of the present day are only beginning to discover," and he adds, truly enough, that physiologists for the last fifty years have attempted "to resolve living functions by physico-chemical means alone rather than from the aspect of man as the recipient of life from a Divine Architect; a designing, discriminating purposeful life, determining special forms and structure." This was Swedenborg's view, for in his *Economy of the Animal Kingdom* (vol. i. § 39) he wrote: "The stupendous machine of the animal body could by no means have come together without a positive directive force; ... and such a directive or formative force is not without but within the chick or embryo." This is the view taken by many eminent naturalists at the present day, such as the veteran A. R. Wallace, who have become convinced of the utter inadequacy of a purely mechanical and materialistic theory of evolution.[2]

I must pass over other contributions made by Swedenborg to physiology and psychology, and will only refer to his remarkable psychological insight concerning the true function of the organs of sensation. The popular view, even at the present day, confounds the perception of sight, hearing, touch, taste, and smell with the organ of sight, hearing, etc. As the philosopher Reid remarks in his *Inquiry into the Human Mind*: "We invariably confound the organs of perception with the being that perceives; the eye is not that which sees, it is only the organ by which we see: the ear is not that which hears, but the organ by which we hear, and so of the rest." Now, these are almost the exact words used by Swedenborg, and published in 1763, the year before Reid had written them. We do not look at the tiny image formed on our retina by external objects, as the photographer looks at the image on the ground glass screen of his camera. The retinal impression is transmitted to the brain through the optic nerve, and a certain tract of brain cells is thereby stimulated; the multitude of separate stimuli thus created, our ego collects into a coherent whole, and we then mentally project outside ourselves a phantasm of the object seen. The combined image we see is a creative act of our own spirit, and leads us to think that the appearance, or phantasm, resembles the thing we are looking at. Now Swedenborg recognized this when he said: "The sight of the eye, strictly speaking, is nothing but the sight of the Spirit produced outwards."

The interesting part of Swedenborg's anticipation of the Scotch philosopher Reid, and of our modern view, is that it occurs in his two works, the *Arcana Coelestia* and in *The Divine Love and Wisdom,* written during the second or spiritual epoch of his life, to which we must now turn.

Swedenborg As A Seer

The Spiritual Significance of Nature

Though his training in the natural sciences during the first fifty years of his life had, as Swedenborg affirms, prepared him for the period of seership, yet from the time of his spiritual quickening onwards, he gave up active scientific investigation and retired from his post as Assessor of the Board of Mines, in order to devote himself to

2 See my Essay on *Creative Thought*, published by Watkins, uniform with this paper.

the higher mission to which he believed he had been called.

But unlike most mystics, he carries into all his writings of this second period the calm, unimpassioned air of science. He is always serene and confident, and knows no wavering in the Divine vocation which he claims has been conferred upon him. For nearly thirty years, he asserts he held daily converse with angels and with many of the departed spirits from earth. He tells us that he recounts what he has actually seen and heard, not mere imagination or ecstasy. Yet we find the clear-sighted philosopher and savant co-existing with the mystic, to whom the spiritual world seemed more real and open to view than the world of sense and outward things.

To the superficial view of most men, he had become mad; to the psychologist, the visions and voices were hallucinations, a secondary self, or an alternating personality. Nevertheless, Mr. F. W. H. Myers states in his great work on *Human Personality* that this period of Swedenborg's life—"one of the strangest lives yet lived by mortal men"—is corroborative rather than destructive of the slowly rising fabric of knowledge, of which he was the uniquely gifted precursor. By this, Mr. Myers means that the results of modern critical investigation in psychical research and the glimpses apparently afforded of the state of life after death certainly confirm the long prior statements made by Swedenborg.

That Swedenborg exhibited, from time to time, supernormal knowledge of earthly things appears absolutely incontestable. The best-known cases are those which the philosopher Kant investigated soon after their occurrence and which he minutely described in 1763.[3] Kant regarded them as proved and says concerning one of them, that "it places the assertion respecting Swedenborg's extraordinary gift beyond the possibility of doubt." This particular case was the statement made by Swedenborg to certain well-known people in Gothenburg that a great fire had broken out that afternoon in Stockholm, some two hundred and fifty miles distant as the crow flies: Swedenborg described with alarm the course of the fire, as it was then raging, and its ravages, destroying the house of one of Swedenborg's friends and threatening his own house, and finally the exact place and time of its termination. This was long before the days of telegraphs and

3 See Kant's letter to Fräulein von Knobloch in Appendix II. to the English edition of *Dreams of a Spirit Seer*, published by Swan Sonnenschein & Co., and Dr Tafel's discussion of the evidence in his *Documents concerning Swedenborg.*

telephones, but two days later a messenger from Stockholm arrived bringing news of this fire just as Swedenborg had described it, and the following day a Royal Courier came from Stockholm who confirmed in exact detail all the particulars Swedenborg had given.[4]

But it is not upon such evidence that the claim of Swedenborg to a higher spiritual illumination can be based. For many evidential cases of clairvoyance (or *telosthesia*, as it has been termed), and of other supernatural gifts, are to be found in the records of psychical research, existing among persons many of whom have no true spiritual life or insight. In fact, religious faith can never be a matter of scientific demonstration, nor be derived from mysterious phenomena; it can only spring from the influx of the Divine spirit to the heart and life of those willing to receive it.

Moreover, Swedenborg himself regarded his amazing experiences as subordinate to the mission to which he believed he had been called. That mission was to expound the inner or spiritual sense of the Word of God, whether that Word is expressed in the inspired writings or in the evolutionary processes of nature. For all nature, all human wisdom and love, is but the progressive revelation and inspiration of the thought of God, unfolding and expressing itself in the worlds and lives around us. This, however, was not the view then held by theologians and the general public, who considered that some few thousand years ago, at the utmost, nothingness was turned into nature in six days, and that both nature and man, after the creative fiat of the Almighty, were left just as we now see them. In fine, orthodox, scientific, and philosophic thought in Swedenborg's day was not allowed to travel outside the literal interpretation of the Bible. Swedenborg, however, declared that just as this earthly human body is but a tabernacle, a clothing of the spirit which survives when the body perishes, so the literal sense of the inspired writings is but a clothing for the spiritual sense, or soul of the Word, which survives when the natural sense, the mere outward body of the Word, perishes. The outward or apparent sense was, however, necessary in the childhood of the race as an introduction to the higher spiritual meaning it enshrines.

And so, also, as regards the phenomenal world around us. Swedenborg asserted that the order and intelligibility of the natural world are derived from the fact that it is the material symbol of a spiritual

4 See my little book on *Psychical Research* (p. 154 *et seq.*) in the Home University Series, where a summary is given of the cases of Swedenborg's supernormal vision.

world, that nature proceeds from, and is at every moment sustained by, the inflow of the Divine, Inscrutable Power. To the pure materialist, this world is self-sustained and has no deeper meaning than the appearance it presents to our senses. It is to such just what a letter appears to those who cannot read—curious, orderly, black strokes on paper, with no further or intelligible meaning. But to those who can interpret written or telegraphic symbols, they give an intelligible message; they put them into touch with the thought and purpose of an unseen personality. And the reason is because the intelligence of the distant writer or operator is related to our intelligence, and hence by symbols, he can convey his ideas to those who have learned their meaning. And the reason why the material world is intelligible, why we can interpret its symbols, and see the unity of its design, is because there must be Intelligence, Thought, and Purpose behind, and ever present in, the phenomena of nature. In a word, an Infinite, Unseen Mind exists, to which our mind is related and from which it proceeds.

But even the materialist must admit that science has shown the visible universe to be a *cosmos* and not a *chaos*; that amidst the mutability of all things there is no capriciousness, no disorder; that in the interpretation of nature, however obscure and entangled the phenomena may be, we shall never be put to intellectual confusion. The magnificent procession of phenomena in the midst of which we live, the realms and abysses of space too vast for the mind to grasp, the unseen molecules and movements too minute and too rapid for the mind to conceive—all these are marching to the music of a Divine order, of an Eternal purpose. On this system of the orderly government of the world and of the intelligibility of nature, our faith in a Supreme Being is rooted, and the progress of modern science, so far from destroying, has confirmed that faith in every thoughtful and reverent mind.

If, then, the natural or phenomenal world be the symbol and effect of a more real and spiritual world, just as language is the symbol and effect of thought, there is a *correspondence,* a mutual relation, between the two. The mutual relation between thought and language, between the mind and the body, between the will and voluntary movement, is a type of the relation between the spiritual and natural worlds. For the effects we notice, such as speaking, walking, etc., are due to directive guidance proceeding from the mind.

Now Swedenborg, in his doctrine of correspondence, declared that there was just such a relation as this between the natural and spiritual universe; that the former is, as it were, the body, which is animated, dominated, and delineated by the spirit. We see daily that no Divine truth, no evil thought, can dwell in any heart without the countenance of the man in whom it dwells bearing silent witness of the fact. As the poet Spenser wrote more than three hundred years ago:

> "For of the soule the bodie forme doth take,
> For soule is forme, and doth the bodie make."

And Swedenborg says: "Goodness and love mould the form into their own image, and cause the joy and beauty of love to shine forth from every part of the face." Here, as elsewhere, we find a connection and correspondence, a relation of cause and effect, between spirit and matter.

In the use of the term *correspondence,* Swedenborg does not mean a mere resemblance or analogy between one object in nature and another, without any necessary relationship between the object and its analogue. He intends the word to signify the actual relation which subsists between cause and effect, between spirit and nature, between the interior and the exterior. The exterior form perishes when separated from the interior and essential cause, just as the body dies when its life has departed, or as nature would cease to exist if the Divine spirit were withdrawn. Swedenborg believed that the doctrine of correspondence was known to an ancient, prehistoric Church, and that the Israelitish Church embodied this doctrine in its "sacrifices, burnt offerings, meat offerings, and drink offerings, with all the particulars belonging to each." These he asserts were correspondences, representations of spiritual truth.

Moreover, Swedenborg assures us that the various objects in the natural world are all counterparts and effects of things and causes in the more real, spiritual world; nay more, that the former exists and subsists from the Divine will through the medium of the latter. Thus he tells us that as life in this world is dependent on the Sun, in like manner all things which exist in the spiritual world depend on the Love and Wisdom which proceed from the Lord; and this Divine proceeding appears as a Sun in the spiritual world, from whom, and in whom, all things in earth or Heaven "live and move and have their being."

This correspondence becomes the more remarkable in the light of our present knowledge. We cannot see, nor can we have any conception of, the real physical sun *in itself* which lies behind the appearance it presents to our senses; but we know it reveals itself to our sight, and gives color and beauty to the world by the visible rays it sheds upon all things. But accompanying the visible sunshine there proceeds from the sun a vast invisible radiation; which, by its warmth and actinic power, gives energy to wind and water, quickens and fructifies the seed, and sustains the mystery of life. And so in like manner the one ineffable God, "whom no man hath seen nor can see," reveals Himself to us by the Lord Jesus, who sheds on the world His Divine Light and Wisdom, which cannot be received into the hearts and lives of men until they are quickened, regenerated, and energized by the Divine Love and Goodness, flowing in invisible streams from the Lord, the Spiritual Sun.

As the Divine Love and Wisdom are translated into their correspondence of heat and light in this world, so *mental states* in the spiritual universe have their correspondence to *spaces,* and the progression of these mental states their correspondence to *times*, in the physical universe. Hence thoughts and affections give rise to the appearance of time and space in the spiritual world. "In the other life," Swedenborg says, "all things there appear as if they were in space, and succeed one another as if they were in time, but in themselves these are changes of state, for this is their source." It is interesting to compare this with Kant's *Dissertation on the Two Worlds* published subsequently.

The spatial distinctions in the spiritual world appear as real and objective as they do here, so that the spirit lives in a world other than itself, a world of distinct personalities existing in apparent time and space. The soul moves, as it were, through space, but really to a *state* where it can associate with others who think and feel alike.

The beautiful conception of Dante, wherein he represents himself as ascending to the higher heavens, not by translation through space, but by seeing his Beatrice grow more and more lovely, is, with Swedenborg, the very law of the spiritual world.[5] The more of goodness

5 See the chapter on Swedenborg in vol. ii. of Vaughan's delightful work *Hours with the Mystics.*

and truth there is in the soul, the more we discern the loveliness and attractiveness of the true and the good; whereas the more of evil in our hearts, the more we long for and gravitate towards all the repulsive forms of evil. And thus, according to our seer, arise the heavens and hells—both are states of infinite variety, combined in one stupendous whole, which, as regards the heavens, Swedenborg calls *Maximus Homo,* the transcendent expression of the Divine Man. As the infinite multitude of cells in a human body are each instinct with life and special use, yet together constitute the one man, so in the vast aggregate of redeemed humanity each life fulfills its allotted purpose in the majestic evolution of the *Maximus Homo.*

Closely connected with the law of correspondence is Swedenborg's doctrine of *degrees.* To this I can only briefly refer, though it enters very largely into all his writings. He states there are two kinds of degrees, which he terms *continuous* and *discrete.* By the former he means a gradually extending or decreasing series, such as the diminution of light or heat with distance from the source, variations in density, etc.; whereas by discrete, or non-continuous degrees, he means things prior and posterior, such as cause and effect, the spiritual and the natural. Continuous degrees are, as it were, variations of length and breadth; discrete degrees, variations of altitude. In the greatest to the least of all created things, he asserts, both these kinds of degrees are to be found. Each kind forms a plane of being by itself, and can never become the other by any increase or decrease of intensity, any more than matter can become spirit by a process of rarefaction. The connection between discrete degrees is to be found in the law of correspondence, according to which each thing in nature corresponds to something in a co-existing spiritual world: the world of causes which produces the world of effects.

Passing to the life of man, Swedenborg asserts there are in him three discrete degrees: the natural, the spiritual, and the celestial. When a man is born, he first comes into the natural degree and gradually acquires understanding and reason by continuous degrees; but the spiritual degree is not opened to him until the love of self is added to love towards others, and this degree advances continuously by the love of spiritual truth and goodness to its summit. Nevertheless, the highest or celestial degree is not opened by these truths, but only by love to the Lord and translating into life the precepts of His Word.

When this occurs, intellectual apprehension is replaced by spiritual perception; in humility of spirit, love of self is lost in a greater love to others, and the soul becomes more and more closely conjoined to God in the spontaneity and fullness of its love to all men. The Divine power thus ascends in man through these three degrees, which may be said to be intellectual, moral, and spiritual enlightenment. Creation, Swedenborg taught, is a receding of the spheres of life emanating from the Creator to the creature, and into those extremes where existence becomes passive, inert, and resistant, as we find in the inorganic world. God, the supreme love, is ultimate reality, the only real self-subsisting Being in the Universe.

Here we touch upon Swedenborg's system of philosophy, which is based not upon a conceptual view of Nature but upon *love* manifested in life as the only reality. That love is life, he argues, may be demonstrated by eliminating all the aspects and affections of love, when life and its activities will be found to cease. Thus in our search for reality, we find it not in abstract reasoning and conceptions, but in the actual world of living experience; and this experience, in all its richness and variety, is rooted in love. For love creates desire, striving towards an anticipated satisfaction and fulfillment. The first effect of life is thus the thought or perception of an end to be attained (not necessarily a conscious perception), then comes the means or cause to attain this end, and finally, the effect or accomplished result. This series—*end, cause,* and *effect*—Swedenborg reiterates, is the manifestation and development of love. Whether in the behavior of a simple cell or in the most complex form of life, we find these three processes, or stages in the process, of love. The desire or affection leads from feeling to action, and this on to the fulfillment of the desire or end—which Swedenborg calls the "effect," using the term "end" to signify the completion of the first stage.

It is worthy of note, as an acute American thinker, Professor Hite, has pointed out, that in the passage to fulfillment, between the stage of want, desire, longing, striving, and that of satisfaction and achievement, there intervenes the field of ideas, wherein the particular desire has emerged from the immediacy of feeling, and by the act of attention has become an object of effort. An objective world has thus been projected, and in a world which seems other than itself, the desire finds its fulfillment, the effort attains its purpose. Hence, affection, idea, and accomplishment is Swedenborg's doctrine of "end, cause, and effect." Creation is, in fact, the self-projection of the Divine idea flowing from the Divine love through the spiritual into the natural. As Swedenborg says: "There are three things in the Lord, which are the Lord—the Divine of love, the Divine of wisdom, and the Divine of use": and these three are correlated with the three stages of end, cause, and effect.[6]

On the vexed question of the freedom of the will, Swedenborg asserts that the human will is free, for that which distinguishes man from other creatures is his conscience and power of choice between right and wrong. "The universe as an automaton," Dr. F. Sewall well remarks, "acting only under the compulsion of Divine law, would never furnish the response which the Divine creative love demands" in a creature whose choice must be free. Through the abuse of this freedom in the sinner, his inner life no longer governs his thoughts and actions; nevertheless, even in the most depraved, the essential inmost life is still God's life. By shunning evils as sins against God, the process of regeneration, necessary for all men, begins. But this process does not end with death; it goes on to all eternity, apparently from human volition, but really from the influx of the Divine spirit. When the Divine love, which is goodness, is consciously received by man, so that it governs his daily life and is manifested in charity of thought and action, man enters into a state of Divine and delightful order which is heaven. Evil, on the other hand, is a state of disorder, a perpetual hell, into which men condemn themselves. Heaven and hell thus begin on earth. It is true the Bible speaks of God condemning sinners to hell and of His anger, but this is the apparent and not the real sense.

Swedenborg draws a striking distinction between apparent and real truth, illustrating this by the apparent rising and setting of the sun, which is really stationary. In fact, is it not the case that, owing to our limitations, we judge all things, and act accordingly, upon what *appears* true to us; that is, upon what is relatively true, and not upon

---

6 The reader is referred to the able paper on "Ultimate Reality" by Professor L. F. Hite of Cambridge, Mass., U.S.A. (published in the *Transactions of the International Swedenborg Congress,* 1910), for a more complete view of Swedenborg's philosophy, which I have only imperfectly touched upon. There is also an excellent work on Swedenborg's psychology by the Rev. T. W. Gorman, M.A., entitled *Christian Psychology* (Longmans, 1875), which contains many valuable citations from eminent writers in corroboration of Swedenborg's views.

what is absolutely true? For we cannot comprehend the absolute and unconditioned. A painter who argued that color was a mere sensation and only relatively true would lamentably fail if he tried to paint a picture by selecting his pigments from an attempted measure of their wavelengths; and even then he would not have reached the absolute. And so in the moral as in the physical order, free will, evil, etc., may not exist absolutely, but to us, conditioned as we are, they are relatively true; and we should grievously err if we denied our responsibility because free will may have no real existence from an omniscient or absolute point of view.

Turning now to eschatology, Swedenborg, divinely illuminated, as I believe he was, declared that the change we call death is not an abrupt transition from the ordered mystery of life and law we know here to a nebulous and unimaginable realm wholly different from the present; but that the other life is a continuation and development of the present life. That the appearance of things in the unseen will be very much as they are here, so that a man cannot at first realize that he is not still living in his material body. That after death we shall pass into an intermediate state, wherein the spirit remains for a shorter or longer time according to the strength of its ruling love, which determines its future bliss or woe.

Moreover, Swedenborg asserted that the thoughts, acts, and affections of our past life are impressed on our interior memory, or, as we should now say, on our subconscious life, and carried with us into the spiritual world; wherein they are gradually disclosed, so that we shall appear to others as we really are in our inmost affections for good or evil. This stripping off of the pretenses and disguises which may have enwrapped us here constitutes the process of judgment in the spiritual world. Differences of creed, or religious errors, or the affliction of hereditary tendencies to evil, so long as they have not led us to a willful and determined rejection of the monitions of conscience, are outward aspects of the mind, and will disappear under the healthier surroundings and educative influences of a larger and fuller life.

Eventually, all will find themselves in association with those whose ruling love is the same as their own, by the inevitable law of "like to like," which is the gravitation of the spiritual world. The deeper meaning and spiritual significance of our present life and its surroundings will be gradually unveiled. But the reign of law and progress will continue, for varieties in receptive faculties, in goodness and intelligence will remain; together with differences of sex, so that in heaven there will be spiritual marriages of those who are consecrated by a perfect union of aim and affection. Work will not cease, for heaven will be a place of constant activity and industry, but work and worship will be transfused and ennobled with the joy of love to God and our neighbor.[7]

Singularly enough, in the condition of life in the spiritual world, we find Swedenborg anticipating *telepathy*; that is, the transmission of ideas from one mind to another independently of the known channels of sense. He tells us that "in heaven there is a communication of all thoughts, so much so that no one needs to ask another what he thinks"; but that angels can and do speak audibly to each other: "It is one of the wonders of the other life that the thoughts and affections of men and angels are known to one another, so much so that no one needs to ask another what he thinks." Again: "In the other life hearts speak, and not lips." And again: "The speech of spirits among themselves is not one of words but of ideas, such as are those of human thought without the words, and therefore it is the universal of all languages."

The lesson which Swedenborg taught with the confidence of assured conviction was that the law of continuity from the least to the greatest remains unbroken when we pass into the spiritual world. "To Swedenborg," as Mr. F. W. H. Myers truly says, "belongs the first emphatic announcement that this life and the next are morally continuous." There is no place for idle luxury, for monasticism, or for sacerdotalism, in such a scheme of eschatology as this. Swedenborg, in fact, tells us "that a life of charity towards our neighbors, which consists in doing what is just and right in every employment, can only be exercised in general as man is engaged in some employment." Such a life tends more truly heavenward, he says, "than a life of piety without charity." As we open our life here and hereafter to the Divine influx, we become conscious partakers of the Divine life; if we close our life to this influx, God still abides in us, but we are not in God.

To these views enlightened Christian thought at the present day

---

7 For a fuller exposition of Swedenborg's views see a little work entitled *The Spiritual World*, by J. Howard Spalding (F. Warne & Co.). This is a brief, lucid, authoritative, and reverent exposition of Swedenborg's spiritual teaching, and is warmly commended to the inquirer.

is assuredly tending. And whilst I, for one, cannot accept some of Swedenborg's statements in the light of modern Biblical scholarship and of our present scientific knowledge, yet theology and science are both under a great debt to the Swedish seer, who "took philosophy and science from earth to heaven, as Socrates had called down philosophy from heaven to earth."

Now I must bring this lengthy discourse to a close. We have seen that one great value of Swedenborg's teaching is that it produces a more vivid and true realization of the immanence of the spiritual world and of the larger life beyond, a more habitual sense of the indwelling and overruling Providence of God, and a more intelligent interest in the study of the Bible and of Nature; teaching us that the reality and significance of the material world cannot be found in itself, but in its relation to the spiritual, to the omnipotent Mind that creates and transcends the phenomenal world through which for a moment each of us is passing: "The things which are seen are temporal, but the things which are unseen are eternal."

ADDENDUM

In the recently published and excellent life of Swedenborg by Mr. G. Trobridge (F. Warne & Co.), a chapter on the "Personal Characteristics" of the Swedish seer gives a charming picture of Swedenborg's simplicity of life and humility of spirit, and his entire consecration to the Divine mission to which he believed he had been called. Thus, in reference to the motto he had chosen—"God's will be done, I am Thine and not mine"—he writes in his diary: "Pardon my saying I am Thine and not mine; it is God's privilege and not mine to say so. I pray for the grace of being Thine, and of not being left to myself."

Even Swedenborg's contemporary opponents and critics testified to the high esteem in which he was universally held. As one said: "Wherever he was known we find a precious memory"; he was utterly free from envy, hatred, or contempt. When contradicted, he usually kept silence, but when compelled to defend himself, did so briefly and with gentleness, showing great respect to his opponent. In fact, as a contemporary says, he was not only a man of extensive learning and widely traveled, but a polished Christian gentleman, who related his spiritual experiences like an open-hearted child. Singularly methodical and abstemious—his dinner, when at home, being a roll soaked in hot milk—he possessed throughout his life, even in his old age at over fourscore years, the most astonishing physical and mental vigor. His prodigious industry is evident in the vast array of books he wrote and published; and his shrewd insight and ability as a statesman is shown in the speeches, chiefly on finance, which he made in the Upper House of the Swedish Parliament. If "to be sane is to be adjusted to one's environment and capable of coping with the facts around us," then Swedenborg is a type of the highest sanity.

Those unfamiliar with Swedenborg, or who have only been deterred by his extensive and tedious works, often view him, as Carlyle initially did, as "an amiable but inane visionary." However, it is worth noting that some years later, Carlyle acknowledged being rebuked for his hasty judgment, made in ignorance. He remarked, "I have since made some personal acquaintance with the man, read several of his books and what biographies of him could be heard of, and have reflected on the singular appearance he makes in the world, and the notable message he was sent to deliver to his fellow-creatures at that epoch. A man of great and indisputable cultivation, strong mathematical intellect, and the most pious seraphic turn of mind—a man beautiful, lovable, and tragical to me, with many thoughts in him which, when I interpret them for myself, I find to belong to the high and perennial in human thought."[8]

After the humble tribute I have endeavored to pay to the memory of this great savant and seer, it may seem ungracious or presumptuous on my part to differ from the opinions held by many adherents of Swedenborg, who have made a far closer and longer study of his writings than I have. But as I have already stated in the foregoing address, some of Swedenborg's views and assertions seem to me impossible to accept. The Divine authorship and verbal inspiration of the Pentateuch, Psalms, and Apocalypse, which Swedenborg upholds, together with his treatment of St. Paul, I cannot agree with, and are views which few theologians would be found to admit. Nor would scientific men agree with his statements that the moon and planets are the abode of human life as we know it.

In fine, with great deference to my Swedenborgian friends, I cannot accept the inerrancy of all Swedenborg's statements about the Scriptures and the spiritual world. We now know how much the

8 Quoted on p. 324 of Trobridge's *Life of Swedenborg.*

subliminal life and memory unconsciously influence the thoughts and utterances of gifted seers.

Nor can I see the wisdom of forming a separate church to inculcate his teaching. It is quite natural that the early followers of Swedenborg, when they found themselves isolated and misunderstood, should form societies for mutual support and sympathy and for the study of his writings. But the invariable effect of schism from a larger body of our fellow-men, though conscience sometimes renders this imperative, is to accentuate points of difference and to produce narrowness and often intolerance; albeit from this latter the members of the New Church appear to be singularly free.

I hope this brief statement of my own position, which it seemed desirable to make clear, may not give offense to one of the most upright, sincere, and thoughtful bodies of Christians whom it has been my good fortune to know.

Presidential Address to the Society For Psychical Research

By Sir William F. Barrett, F.R.S.
January 29th, 1904.

My first duty is to thank you most heartily for the honor you have conferred upon me today. When I recall the names of the illustrious men who have preceded me in this chair, I can only ask your indulgence for the deficiencies of which I am very conscious and express the hope that our Society may not suffer from having chosen as its President one from a remote distance, and who can lay no claim to the eminence of his predecessors.

It has not been the custom formally to move a vote of thanks to our out-going President, but I am sure you would desire me to voice the grateful appreciation, which we all sincerely feel, of the great services which Sir Oliver Lodge has rendered to the Society during his three years' tenure of office. In his presence I cannot express all I would like to say, but this much perhaps he will permit, that during his Presidency, his influence and high position in the ranks of science have been of inestimable value to the Society, whilst his simple and unaffected courage has been a noble stimulus to us all.

Here too, perhaps, you will allow me to say how much we owe to the ungrudging and arduous labour of our Hon. Sec. Mr. Piddington, and to the skill and care with which our Hon. Treasurer Mr. H. A. Smith has, almost from the foundation of the Society, conducted its finances, and to the legal advice and assistance, in the incorporation of the Society and other matters, which Mr. Smith and Mr. S. C. Scott have freely given to us. Though I am quite sure they would deprecate any allusion to their services, yet I know you will join with me in expressing our most grateful and cordial thanks to each and all of these gentlemen.

This month we enter upon the 23rd year of our existence, and in spite of the irreparable losses we have suffered in the past through the removal from this life of those who seemed to be the very pillars on which the Society rested—Gurney, Sidgwick, Myers—losses which naturally gave rise to many gloomy forebodings—nevertheless we have no reason for despondency. On the contrary, the present flourishing condition of our Society far exceeds that which even I, the most sanguine of its founders, could have anticipated. Both numerically and financially we are stronger than we have ever been; the number of our Members and Associates throughout the United Kingdom, after eliminating all losses through death, etc., is 832; in addition we have 530 Members and Associates in the American Branch, and on the Continent kindred societies are springing up.

As you will see by the printed statement of accounts, our financial condition is also satisfactory. Additionally, thanks to a generous legacy, we have a considerable reserve fund, which we hope our members will increase through their liberality. The Council has placed this reserve fund in the hands of Trustees and has determined to appropriate the interest of it, when it reaches the sum of £8000, to the purpose of experimental research in subjects that fall within the scope of our Society.

It is true that no original research of any value can ever be made to order. The progress of knowledge in psychical science, as elsewhere, must ultimately rest upon the self-sacrificing and intelligent work of those in whom a keen interest for a particular branch of inquiry has been excited, and whose enthusiasm is not stimulated by, nor can it be repaid in, the coin of the realm. However, the investigator has to live, and to enable him to do so while he is conducting any hopeful inquiry, sanctioned by the Council, a modest sum will be appropriated to him as a grant in aid.

We hope to receive applications from competent inquirers of either sex who will undertake special lines of investigation. It might also be desirable, though this is my personal opinion, to follow the examples of older scientific societies and award medals or premiums for any good piece of original work. There is an immense field for investigation; psychical research is almost virgin ground as far as strict scientific inquiry is concerned. Although the worker is not likely in this generation to meet with much encouragement from the world at large, and still less any benefit to himself, he may be sure his toil will not be in vain. I would therefore earnestly appeal to any members or friends of our Society to give us what help they can. If personal assistance is not possible, their contributions to the completion of our research fund will be gratefully welcomed.[1]

And this brings me to a point to which it is desirable to refer. The work of the Society has hitherto been in comparatively few hands, and the center of gravity of the Society inevitably falls in that place (such as Cambridge) where the most active workers are or have been found. This has led some of our members to feel perhaps a little aggrieved that the center is not within their own precincts, that they are not more in touch with the work of the Society, and that matters of great interest may be occupying attention at headquarters and yet no information may reach them. If this represents any general feeling, is not the remedy to be found in wider and more earnest cooperation in our work on the part of the members generally? The *Journal* of the Society was established to encourage this and to be an organ of communication between the members. I would be glad to see it more freely used for this purpose, and for the speedy publication of any first-hand evidence that our members may possess; and also for the prompt communication of information from headquarters which would be of interest to members generally. I hope, therefore, without any departure from the caution and reserve, which are so necessary in dealing with the phenomena that engage our attention, we may be able to have a more frequent interchange of thought and opinion between the Council and the body of our members, and I shall be glad if I can be of any service in promoting this.

As regards wider cooperation in our work, I would venture to suggest that *local groups* should be formed, meeting weekly or monthly for investigation or discussion.[2] If you ask me what work can be done by these local groups, a beginning might be made in the determination of two problems, which though they may be simple and dull, are nevertheless of considerable importance and can only be settled by widespread assistance.

One of these problems may be stated thus:

(1) Assuming the existence of telepathy, is the transference of thought from one person to another independently of the recognized channels of sen-

---

1 In consequence of this appeal two members of the Society wrote to me that they would contribute £50 each if eighteen others would do the same so as to make up £1000 in the present year; £11.3 3s. has since been received, and perhaps some readers may feel moved to contribute towards the completion of the sum.

2 No member of the Society has done more in this direction than Colonel Taylor, R.E., and I hope his example and admirable method of investigation will be widely followed.

sation a faculty in some slight degree possessed *by all*, or is it confined only to a few? Prof. Eichet and others have made experiments on this question, but it is still an open one. It is very easy to devise methods of experiments between two persons, choosing the simplest form, such as tossing a coin, and noting whether the right guesses in, say every 100 trials, somewhat exceed the number which chance alone would indicate. But patience is necessary in all investigations to make the result of any value, and also precautions with which no doubt you are familiar, such as the avoidance of facial or other indications. There are other questions in connection with telepathy which need elucidation to which I will refer presently.

The other problem is:

(2) To what extent is motor-automatism common among mankind? By motor automatism is meant the purposive movement of one's voluntary muscles without any intention or conscious effort on our part. The simplest and most sensitive test is to hold a forked twig in your hand after the manner of a dowser and see if the twig curls up of its own accord when you walk around your garden or go in search of some hidden object. Someone or more members of your investigating circle will probably succeed, in spite of their incredulity. Now it is important to know (1st) the percentage of success among all our friends, and (2nd) whether other phenomena of motor automatism occur with the same individual, such as the movement of the pendule explorateur, automatic writing, etc., and (3rd) whether these movements are merely fortuitous, or whether they are veridical, that is to say, do they give us truth-telling information of something the individual could not have ascertained by the use of his ordinary perceptive powers?

Now these problems require the willing cooperation of a number of intelligent investigators, but they do not require costly apparatus nor any particular training beyond that of careful observation; nor are these experiments in the least injurious to the operator, nor likely to excite any opposition even from a hypersensitive conscience. It is true they may seem trivial and tiresome: if any person thinks so, I am afraid they would think the details of all scientific investigation trivial and tiresome. The work of a bricklayer or a hodman in building a house would seem to be very monotonous if we did not regard the end in view.

I AM sometimes asked what our Society has already achieved, what has it done to justify its existence? The reply to this is found in the eighteen closely printed volumes of our *Proceedings,* and the eleven volumes of our

Journal, containing an immense mass of evidence, the record of carefully sifted observations or of stringent experiments. These form a storehouse of material which we have every reason to believe will become increasingly valuable to students both of psychology and philosophy in the not distant future. Unquestionably, a change of opinion is gradually coming about through the work of our Society. The widespread and unreasoning prejudice which 25 years ago existed against all psychical enquiry is breaking down. This is seen in the list of distinguished men who have become members of our society, and here I desire to welcome one of our great English savants, a man of European reputation, who has recently joined our ranks, and this coincidentally with his election to the high position of joint Hon. Secretary to the Royal Society.

But although there is a more open mind on the part of science towards psychical research, it must be confessed it is still looked at somewhat askance by the leaders and organs of official science. It is worth a moment's attention to consider why this should be. No one asserts that the knowledge we are seeking to obtain is unimportant, for as the learned Dr. Glanville said 200 years ago about similar subjects to those we are studying, "These things relate to our biggest interests; if established they secure some of the outworks of religion." Nor, so far as I know, does anyone assert we are hasty and incautious, or unscientific in our method of investigation. No doubt one reason for the present attitude of official science towards us has been the prevalence and paralyzing influence of a materialistic philosophy, which denies the possibility of mind without a material brain, or of any means of access from other minds to our mind except through the recognized channels of sensation. Both these propositions are of course denied by our religious teachers, who assert that a spiritual world does exist, and that the inspired writings were given supersensuously to man. Nevertheless, as a body, though with some notable exceptions, even *they* do not welcome us with open arms. The common ground and official view of both science and religion is that all extension to our existing knowledge in their respective departments can only come through the channels recognized by each; in the one case the channel is bounded by the five senses, and in the other case it is that sanctioned by authority. We must all admit that even unconsciously authority has a large share in molding our convictions and determining our conduct, in fact we cannot emancipate ourselves from its subtle influence. As a rule this is beneficial, unless it can be shown that authority is untrustworthy; but the attempt to prove that it is so is sure to be an ungracious and difficult task, and almost certain to bring odium to bear upon those who, if they eventually prove to be right, are in a subsequent generation hailed as benefactors of the race.

Some years ago, the late Professor von Helmholtz, a highly learned man, visited Dublin. At that time, I had recently published a paper presenting the first *prima facie* evidence of a new scientific concept called thought-transference, now known as telepathy. Helmholtz, a renowned physiologist and physicist, engaged in a conversation with me on the subject and concluded by saying: "I cannot believe it. Neither the testimony of all the Fellows of the Royal Society, nor even the evidence of my own senses, would lead me to believe in the transmission of thought from one person to another independently of the recognized channels of sensation. It is clearly impossible." The respect that is due to so great a man renders it necessary to show in a few words why this statement (one that used to be common enough) is wholly indefensible. First, the phenomena in question, and all those within the scope of our Society, are not *contradictions* but merely extensions of our existing knowledge. They may be strange and inexplicable, but that only suggests that the evidence supporting the new facts must be recognized as *adequate*. As Laplace long ago stated in his *Theory of Probabilities*: "We are so far from knowing all the agents of nature, and their various modes of action, that it would not be philosophical to deny any phenomena merely because in the actual state of our knowledge they are inexplicable. This only ought we to do—in proportion to the difficulty there seems to be in admitting the facts should be the scrupulous attention we bestow on their examination."[3] This is the true spirit, as demonstrated by recent discoveries concerning Radium. These facts seemed to contradict some of our previous knowledge. We always thought of an atom, as Lucretius did, "strong in solid singleness," as the most immutable and immortal thing in the physical universe. Now it appears to be capable of disintegration and transmutation, and the views of the alchemists are beginning to revive: soon we may be looking for the "philosopher's stone"—the substance that by its presence enables the transmutation of other heavy atoms to occur. Thus does the whirligig of time bring its revenges.

But to return. There is another fallacy in the scientific view expressed by Helmholtz. He said, as many do, that nothing could make him believe in such phenomena. But belief is not a voluntary act of the mind; it cannot be given or withheld at pleasure. It is, obviously, an involuntary state that follows if our judgment considers the evidence adduced both adequate and conclusive. We can, of course, as many do, refuse to listen to the evidence. It is worth noticing that in all our minds there is a tendency to repel the

3 Laplace, *Théorie Analytique des Probabilités,* Introd., p. 43.

intrusion of any ideas unrelated to our usual habits of thought, which therefore involve an uncomfortable dislocation of our mind. So, attention to evidence of this character is a difficult act of self-conquest. Hence, every new departure in thought has to encounter great mental inertia, and wisely so, as it prevents hasty and foolish aberrations of mind. But when attention is given, and the evidence considered adequate, it is sheer nonsense to say you won't believe it.

Is there any other reason why science should not ungrudgingly recognize the evidence so amply given in our Proceedings? I have recently inquired among some of my scientific friends who stand aloof from us to understand their reason for doing so. Of course, life is short, the demands of each particular branch of scientific investigation become increasingly exacting, and few have time to consider the evidence. That is obvious, but why do they shrug their shoulders when you mention, say, telepathy or the faculty of dowsing? Their attitude reminds me of an anecdote told by that remarkable woman, Miss Caroline Fox, which I think is mentioned in the memorials of her life. The charming residence of Miss Fox in Cornwall was the meeting ground of many famous men of the last generation. On one occasion, that great Irishman, Sir W. Eowan Hamilton, met Sir G. Airy, the then Astronomer-Royal, there. Hamilton had just published his famous mathematical discovery of quaternions and was, I believe, explaining it to Airy. After a short time, Airy said, "I cannot see it at all." Hamilton replied, "I have been investigating the matter closely for many months, and I am certain of its truth." "Oh," rejoined Airy, "I have been thinking over it for the last two or three minutes, and there IS nothing in it." This is why some of our scientific friends shrug their shoulders at our researches. They feel competent, after a few minutes' consideration, to reject conclusions which may have cost us years of investigation.

In fact, nine-tenths of the positive opinions we are accustomed to hear about psychical research are given judicially. That is, the objector speaks of his conclusions as positively as if it were his office to know the truth, and implies that any opposition is a thing for him to judge. "He is annihilated," as Professor De Morgan pointed out some time ago, "by being reduced, no matter how courteously, from judge to counsel. But this is what must be done. The jurisdiction must be denied. The great art is not to pull him off the bench without ceremony, but to pull the bench from under him, without his exactly seeing how he came to tumble, and without proceeding to sit upon it yourself."

Enquiry among my scientific friends has shown me that the root of much, perhaps most, of the scientific skepticism towards our work is not because the phenomena are startling or inexplicable, but because they cannot be repeated at pleasure; hence so very few scientific men have the opportunity of verifying the observations some of us have made. They do not doubt our good faith, but they think we may have been mistaken in our conclusions, and until we can reproduce the phenomena before them, they feel justified in distrusting our results. This might well give ground for suspense of judgment, but surely not for any hostile attitude. It is, of course, most desirable to be able to repeat our experiments at pleasure, but the very nature of our enquiry precludes this. We do not refuse to believe in the fall of meteoric stones unless we can see one falling. We may require a good deal of well-attested evidence for their fall, but, once the fact is established, the stringency of the evidence demanded immediately relaxes. Now, unquestionably there are at present more capable witnesses who can speak from personal and careful enquiry as to the fact of telepathy, or of what is called spiritualistic phenomena, than there are persons living who can testify to having seen the actual fall from space of meteoric stones.

The fact is, our scientific friends do not realize the profound difference that exists between the conditions of a physical and a psychical experiment. We know what conditions are requisite in the former case; we do *not* know what they are in the latter, and hence the difficulty of all psychical investigation and the uncertainty of the reproduction of any given phenomenon.

A moment's consideration shows that the demand made upon us by science for the demonstration at any moment of a particular psychical phenomenon is inconsistent with the very object of our enquiry. Psychical experiments depend on the mental state of the subject; you may tell a person to do something, but whether he does it or not depends on the person addressed. Physical experiments are independent of our volition; a magnet attracts iron, or sets itself in the magnetic meridian, irrespective of our mental condition. This obvious difference between the two sets of phenomena is constantly overlooked. Physical science excludes from its survey the element of personality, with which we have to deal and over which we have little or no control. It regards all phenomena as strictly impersonal, and finds abundant field for investigation within the narrow limits it has marked out for itself: these things it regards as real, the rest as shadowy. The truth is, of course, exactly the reverse. The reality of which we are conscious is our self, our personality. It is the phenomena of external nature which are shadowy; shadows cast by some reality of which our senses tell us absolutely nothing.

There is, however, no reason why the methods so successfully pursued by science should not also be pursued in the study of the complex and shifting phenomena of human personality. Now this is precisely the object of our Society—the accurate investigation of that wide range of obscure but wonderful powers included within the mysterious thing we call ourselves. Albeit we are but at the beginning of a task so vast that it may, in time to come, make all the discoveries of physical science seem trivial, all its labours seem insignificant in comparison with the stupendous problems that are before us.

We need, therefore, much more experimental evidence in every department of our work. As far back as 1876, in a paper read before the British Association in that year, I stated that before science could attack with any hope of success the investigation of alleged spiritualistic phenomena, we must know whether definite ideas can unconsciously be communicated from one person to another: whether such a thing as thought-transference does really exist. Evidence was adduced in favor of this hypothesis. We have done much since then, but much remains to be done before telepathy can take its place as an accepted axiom of scientific knowledge.

I REFERRED at the beginning of my address to some problems in connection with telepathy that await solution. Permit me for a few moments to return to this.

There is one question in regard to telepathy and similar psychical phenomena, which is likely to remain an outstanding difficulty. By what process can one mind affect another at a distance? Physical science teaches us that there is no such thing as "action at a distance." Energy at a distance reaches us either by the translation of matter through space, like a flying bullet, which carries the energy; or by the intermediary action of some medium, like the transmission of sound-bearing waves through the air, or of luminiferous waves through the ether, the energy being handed on from wave to wave. We may talk of brain waves, but that is only unscientific talk; we know of nothing of the kind. Neither do we know how gravitation acts across space; by what means such tremendous forces as bind the solar system together are either exerted or transmitted, we know absolutely nothing. We don't talk of gravitation waves; we wait for further knowledge on this mysterious problem; and in like manner, we must patiently wait for more light on the mode of transmission of thought through space. It may well be that thought transcends both matter and space and has no relation to either. That mass, space, and time may only be but the mental symbols we form of our present material system and have no ultimate reality in themselves.

Another question is as follows: May not the uncertainty and difficulty of our experiments in thought-transference partly arise from the fact that we are not going to work the right way? We try to obtain evidence of the transmission of a word or idea through some conscious and voluntary act on the part of the percipient. We wait for a verbal or written response. Is not this a mistake? Ought we not rather to seek for evidence of thought-transference in the region of the subconscious life? I believe in the case of both the agent and percipient, the conscious will plays only a secondary part. This is also true, I think, in all cases of suggestion and of the therapeutic effect of suggestion. It is notably seen in the cures wrought by what is known as Christian Science. I happen to have had occasion to study these somewhat carefully of late, and undoubtedly remarkable cures are effected, it may be by suggestion, but without the usual suggestive treatment; the only formula is the "Allness of God" and the "non-existence of disease." But the healing processes are set going by a purely subconscious act. And so in telepathy, we need to hand over the whole matter to the subliminal activities. The difficulty is how to do this. Hypnosis is one way. And in the ordinary waking state, the agent, who makes the suggestion or transmits the idea, would, I believe, do so more effectively if, after the intention had soaked into his mind, he left it alone, so far as any conscious effort was concerned. And the percipient should be as passive as possible, make no effort to guess the word, but allow the perception to reveal itself through some involuntary action. Automatic writing would be the most effective, but that is not very common; the twisting of the forked dowsing twig might be utilized, indicating the letters of the alphabet by its motion; or in other ways. In the historical researches I have made on the so-called divining rod, I found it was used in this very manner two centuries ago. In fact, what we need to learn is the language of the subliminal life, how it speaks to us, how we can speak to it. The voluntary action of the muscles in speech or gesture is the language of our conscious life; the involuntary action of our muscles and emotional disturbance appear to be the language of the subconscious life.

Another point to consider is the frequent *delay* in the impression perceived by the individual. I observed this repeatedly in my initial experiments with thought-transference 25 years ago. Sometimes, the correct response to a previous experiment would come in response to a later and different experiment. I have also noticed this in dowsing, where, with some dowsers, the movement of the twig lags behind the moment of the impression; it turns *after* the dowser has passed slightly beyond the correct spot. We have similar phenomena in physical science. The magnetic state of iron lags slightly behind the magnetizing force it is subjected to; this is known as *hysteresis,* from a Greek word meaning to lag behind. Therefore, I believe there is both a psychical and a physical hysteresis, and if so, it should be considered in our experiments. It is unlikely that any psychical action, even telepathy, occurs without some preceding change in the nerve tissues; in technical terms, neurosis must always precede psychosis. This change must then rise until it is of sufficient magnitude to create the reflex that moves the muscles. All this involves time, which may vary, and thus account for the occasional lag we observe.

Other questions suggest themselves. Is it the idea or the word, the emotion or the expression of the emotion, that is transmitted in telepathy? Probably the idea. If so, it affords a hint towards the interchange of thought amongst the race in spite of differences of language. Language is but a clumsy instrument of thought, and quite incommensurate to it; its arbitrary signs show it to be but the rudiments of a system which the evolutionary progress of the race may lead us to hope will be more perfect in the future. How much more accurately should we be able to transmit complex ideas and subtle emotions if thought could evoke thought without the mechanism of speech? This may now be the case in the state of life in the unseen. The sanctity and privacy of our minds will, however, require to be protected from unwelcome intrusion, and this, so far as our conscious life is concerned, will doubtless be within our own power to effect, so long as we retain control over our self-hood, our true personality.[4]

Then, again, may not animals share with man this telepathic power? They have in some directions keener perceptive faculties than man, and there is evidence that they are strongly affected by what we call apparitions. It may be that animals, and insects like the ant and the bee, do communicate with each other by some process analogous to telepathy. It is worth trying to find out whether, say, a favorite dog can respond to a telepathic impact from his master. In centuries to come, it is just possible that through some such interchange of feelings, we may get into closer communion with all

4 In that remarkable book published some 70 years ago, Isaac Taylor's *Physical Theory of Another Life*, Chap. viii., will be found a prevision of telepathy and of some of the ideas contained in the foregoing paragraph.

sentient things.

There is one argument in favor of the existence of something analogous to thought-transference, which—so far as I know—has not been used. I believe it is a legitimate argument because it is based on the underlying unity that exists throughout Nature. The theory of gravitation teaches us that every grain of sand on every seashore in this world, every particle of salt in every salt cellar, is forever pulling every grain of sand or salt, not only on this earth but on every planet or star in the whole Universe. And vice versa, for there is a reciprocal influence ever going on between these myriads of remote things. Moreover, such is the solidarity of the Universe that an interchange of radiation, as well as attraction, is ever taking place between things on this earth and also between our planet and every member of the solar system. No fact in physical science is more certain than this. May not this "theory of exchanges," this mobile equilibrium, extend to the psychical as well as the physical universe? Tennyson, with poetic prescience, asks in *Aylmer's Field:*

"Star to star vibrates light, may soul to soul
Strike thro' a finer element of her own?"

Certainly it seems very probable that every centre of consciousness is likely to react telepathically upon every other center.[5]

It is hard to believe that the play of *vital* forces should be more restricted than that of the *physical* forces; that radioactivities should be confined to inanimate matter. If this unconscious radiation and reaction is going on between mind and mind, then observed cases of telepathy would simply mean

5 Since this address was delivered my attention has been drawn to Mrs. Browning's striking sonnet on "Life," wherein the same idea is elaborated; poets are certainly wonderful pioneers of thought; before telepathy was thought of Mrs. Browning wrote:

"Each creature holds an insular point in space;
Yet what man stirs a finger, breathes a sound.
But all the multitudinous beings round
In all the countless worlds, with time and place
For their conditions, down to the central base,
Thrill, haply, in vibration and rebound,
Life answering life across the vast profound,
In full antiphouy, by a common grace ?
I think this sudden joyaunce which illumes
A child's mouth sleeping, unaware may run
From some soul newly loosened from earth's tombs:
I think this passionate sigh, which half begun
I stifle back, may reach and stir the plumes
Of God's calm angel standing in the sun."

the awakening of consciousness to the fact in certain minds. Why some and not all minds, and why so fitfully the conscious perception should be aroused, are problems we must leave to the future; they are quite consistent with what we find everywhere in nature. For my own part, I am disposed to think this interchange is common to the race and is the chief reason why all men are insensibly molded by their environment. Only, as I said just now, I believe the telepathic exchange emanates from and affects the subconscious part of our personality. It is potentially conscious and may, and probably will eventually become an integral part of our self-consciousness.

We know as a matter of fact that a vast number of impressions are constantly being made upon us, of which we take no heed; they do not interest us, or they are not strong enough to arouse consciousness. But the impressions are *there*, they leave a mark upon us though we are not aware of it, and they may float to the surface or be evoked at some future time. One of the most certain and striking results of the investigations made by our Society is that the content of our subconscious life is far greater than that of our conscious life. Our minds are like a photographic plate, sensitive to all sorts of impressions, but our ego develops only a few of these impressions; these are our conscious perceptions, the rest are latent, awaiting development, which may come in sleep, hypnosis, or trance, or by the shock of death, or after death.

But even here and now, this subconscious radioactivity of thought may already play some part in the growing sense of sympathy and humanity we find in the race. And what a change would be wrought if it were suddenly to become an element of consciousness among mankind. To realize the brotherhood of the race would not then be a pious aspiration or a strenuous effort, but *the* reality of all others most vividly before us; involuntarily sharers in one another's pleasures and pains, the welfare of our fellow men would be *the* factor in our lives which would dominate all our conduct. What would be the use of a luxurious club and Parisian cooks if the privation and suffering of the destitute were telepathically part and parcel of our lives? Slowly the race *does* seem to be awakening to the sense of a larger self, which embraces the many in the One, to

"A heart that beats
In all its pulses with the common heart
Of humankind, which the same things make glad,
The same make sorry."

The instinct of true religion, like the insight of the true poet, arrives at some great verity without the process of reasoning or the need for proof. This has been the case with the belief in prayer and its efficacy. Skepticism scoffs at the mystery of the direct action of mind on mind and the even greater mystery of the movement of the Infinite by the finite—but faith remains unshaken. For us wayfaring men, however, reason needs some help in climbing the heights attained by faith. Is not this help afforded by the steps slowly being cut in the upward path through psychical research? What is telepathy but the proof of the reasonableness of prayer? No longer need our reason rest content with the plausible explanation that prayer can do no more than evoke a subjective response in the suppliant, that it is inconceivable how the Infinite and the finite mind, the One manifest in the many, can have any community of thought. On the contrary, if telepathy is indisputable, if our creaturely minds can, without voice or sensation, impress each other, the Infinite mind is likely to have revealed itself in all ages to responsive human hearts. Some may have the spiritual ear, the open vision, but to all of us, there comes at times the echo of that larger Life which is slowly expressing itself in humanity as the ages gradually unfold. In fact, the teaching of science has always been that we are not isolated in, or from, the great Cosmos; the light of suns and stars reaches us, the mysterious force of gravitation binds the whole material universe into an organic whole, the minutest molecule and the most distant orb are bathed in one and the same medium. But surely beyond and above all these material links is the solidarity of Mind. As the essential significance and unity of a honeycomb is not in the cells of wax, but in the common life and purpose of the builders of those cells, so the true significance of nature is not in the material world but in the Mind that gives it meaning, and that underlies and unites, that transcends and creates, the phenomenal world through which for a moment each of us is passing. "The things which are seen are temporal, but the things which are unseen are eternal."

I WILL now turn for a few minutes to another branch of our researches, which has special interest for me as it was this subject that first aroused my interest in experimental psychology and to which I gave many months of experiment long before our Society was founded. I refer to Hypnotism.

There are no doubt many present who remember the outcry that was once raised against the investigation of hypnotism, then called mesmerism.

Constant attacks were made by the medical and scientific world on the one hand, and by the religious world on the other, upon the early workers at this subject. They were denounced as impostors, shunned as pariahs, and unceremoniously pitched out of the synagogues both of science and religion; and this within my own memory. Physiological and medical science can only hang its head in shame when it looks back upon that period. What do we find today—the subject of hypnotism and its therapeutic value recognized! It has now become an integral part of scientific teaching and investigation in several medical schools, more especially on the Continent. I think our Society may fairly lay claim to have contributed to this change of view, and the work of our members, Edmund Gurney, and Doctors Arthur Myers, Milne Bramwell, and Lloyd Tuckey, has added much to the knowledge of a subject the importance of which it is difficult to overrate. It is also worthy of note how the former neglect of this subject by science relegated it to the ignorant and the charlatan, and its practice to mysterious and often mischievous public amusements. These are now less common; and though the public apprehension of the dangerous abuse of hypnotism is grossly exaggerated, for it is less open to abuse than chloroform, I, for one, am strongly of opinion that we, as a Society, should discourage, and (as in many Continental countries) get the legislature to forbid the practice of hypnotism except under proper medical supervision.

Now I think it is the duty of our Society to cherish the memory of these courageous seekers after truth, who were the pioneers in this and other branches of psychical research. The splendid and self-sacrificing labors of those distinguished physicians, Doctors Elliotson and Esdaile, in the fields of hypnotic therapeutics and painless surgery under hypnosis should never be forgotten, any more than the later work of Dr. Braid of Manchester. Dr. Elliotson, though at the head of his profession, sacrificed everything for the advancement of this branch of knowledge. The mesmeric hospital in London, and the similar hospital founded in Calcutta by an enlightened Governor-General and placed under the care of Dr. Esdaile, did remarkable work, too little known at the present day. I am therefore glad to see in Dr. Milne Bramwell's *magnum opus* on hypnotism that he draws special attention to the labors of Elliotson, Esdaile, and Braid. And it is to be regretted how completely these pioneers are ignored in the works on suggestive therapeutics by Dr. Bernheim, Dr. Liebeault, Dr. Schofield, and some others.

Leaving this part of our subject, now within the purview of science, let us pass to the extreme or advanced wing of psychical research; to that part of our work on which considerable differences of opinion exist even within our Society. I refer to spiritualistic phenomena. With regard to these, we must all agree that indiscriminate condemnation on the one hand, and ignorant credulity on the other, are the two most mischievous elements with which we are confronted in connection with this subject. It is because we, as a Society, feel that in the fearless pursuit of truth, it is the paramount duty of science to lead the way, that the scornful attitude of the scientific world towards even the investigation of these phenomena is so much to be deprecated. Hence, as in the case of those who were the pioneers in the study of hypnotism, we ought not to forget the small band of investigators who before our time had the courage, after patient inquiry, to announce their belief in what, for want of any better theory, they called spiritualistic phenomena. No doubt we can pick holes in their method of investigation, but they were just as honest, just as earnest seekers after truth, as we claim to be, and they deserve more credit than we can lay claim to, for they had to encounter greater opposition and vituperation. The superior person then, as now, smiled at the credulity of those better informed than himself. I suppose we are all apt to fancy our own power of discernment and of sound judgment to be somewhat better than our neighbors'. But after all, is it not the common sense, the care, the patience, and the amount of uninterrupted attention we bestow upon any psychical phenomena we are investigating, that gives value to the opinion at which we arrive, and not the particular cleverness or skepticism of the observer? The lesson we all need to learn is that what even the humblest of men *affirm* from their own experience is always worth listening to, but what even the cleverest of men, in their ignorance, *deny* is never worth a moment's attention.

The acute and powerful intellect of Professor De Morgan, the great exposer of scientific humbug, long ago said, and he had the courage publicly to state, that however much the Spiritualists might be ridiculed, they were undoubtedly on the track that has led to all advancement in knowledge, for they had the *spirit* and *method* of the old times, when paths had to be cut through the uncleared forests in which we can now easily walk.[6] Their *spirit* was that of universal examination unchecked by the fear of being detected in the investigation of nonsense. This was the spirit that animated the Florentine Academicians and the first Fellows of the Royal Society 250 years ago; they set to work to prove all things that they might hold fast to that which was good. And their *method* was that of all scientific research, viz.,

6 See Preface of *From Matterr to Spirit*, p. xviii.

to start a theory and see how it worked. Without a theory, "facts are a mob, not an army." [Meteorology at the present moment is buried under a vast mob of observations for want of ingenuity in devising theories; *any* working hypothesis is better than none at all. And so I agree with De Morgan that the most sane and scientific method in psychical research is not to be afraid of propounding a theory because it may seem extraordinary, but have courage to do so and see if it works.] The theory of thought-transference led to the accumulation of evidence which bids fair, sooner or later, to place telepathy among the established truths of science.

The amusing feature in the progress of knowledge is that, usually, critics who resist as long as they can a new theory are apt afterwards, when the theory becomes widely accepted, to use it indiscriminately, as if it covered all obscure phenomena; and so it becomes a kind of fetish in their thoughts. We are all familiar with the imposture theory, with the coincidence theory, and with the telepathic theory; each excellent in their way, but most foolish and unscientific if we allow any one of them to obscure our vision or paralyze our investigation. What is to be reprobated, as De Morgan said, "is not the wariness which widens and lengthens enquiry, but the *assumption* which prevents and narrows it."

Instances are well known of the most acute and careful enquirers, trained psychical detectives we might call them, who having begun with *a priori* reasoning and resolute skepticism, when they have thrown aside their preconceived assumptions, and given the necessary time and patience to the investigation of one particular case, have gone over to the spiritualistic camp. They may be right or wrong in their present opinion, but we must all admit they have far better reasons for forming a judgment than any of us can have. If they are right, it follows that the particular case they have investigated is not likely to be a solitary one, but typical of similar cases with us as well as with them.

Pray do not suppose I hold a brief on behalf of spiritualism either as a practice or a religion. On the contrary, to my mind, few things are more dismal than the common run of spiritualistic seances. Sometimes they revolt one's feelings, and always they are a weariness to the flesh. Perhaps the manifold experiences I have had have been unfortunate, and I freely admit my remarks apply more particularly to sittings with professional mediums, where what are called physical manifestations take place, which always seem to be on a lower plane, even where the possibility of fraud has been carefully excluded. Nevertheless, if we can get at truth, what does it matter whether we draw it from a well or drag it from a bog?

It is impossible, however, not to feel some sympathy with the common objection of the doubter that the phenomena are of so paltry a character. But we cannot prescribe to nature; we cannot get rid of the leprosy of doubt by choosing rivers of our own to wash in. And so we must be content with what we find. After all, from a scientific point of view, *nothing* can be paltry or mean that manifests *life*.

Bacteriologists spend their days searching for evidence of the lowest forms of life. And surely any evidence of personality that gives us the faintest, rudest sign that life still persists though the clothing of the body be gone is worth infinite trouble to attain. Though it may be

> "Only a signal shown and a voice from out of the darkness,"

it is not paltry. In fine, it is this natural human longing that renders a dispassionate consideration of the facts, a calm and critical weighing of the evidence, so difficult and yet so imperative.

We must, however, bear in mind, as was pointed out by the present Prime Minister in the remarkable address he delivered from this chair, that if science had first attempted to include in its survey not only physical but psychical phenomena, it might for centuries have lost itself in dark and difficult regions, and the work of science today would then have been *less*, not more complete.[7] This is very true; the foundations of our faith in the undeviating order of nature had to be laid by the investigation of the laws of matter and motion and by the discovery of the orderly evolution of life. What science has now established is that the universe is a cosmos, not a chaos; that amidst the mutability of all things, there is no capriciousness, no disorder; that in the interpretation of nature, however entangled or obscure the phenomena may be, we shall never be put to intellectual confusion.

Now, if instead of investigating the normal phenomena of the world in which we live, science had first grappled with supernormal phenomena, it would not have reached so soon its present assured belief in a reign of law. We believe that fuller knowledge of the obscure phenomena we are investigating will in time come to us, as it has in other branches of science, but the appearances are so elusive, the causes so complex, the results of work sometimes so disheartening, that we need the steadying influence of the habit of thought engendered by science to enable us patiently and hopefully to pursue our way.

7 *Procetdings* S.P.R., vol. x., p. 5.

Possibly historical research among the most ancient records may give us fragments of unsuspected information; for it is very probable that many, if not all, the psychical phenomena we are now investigating were known, and the knowledge jealously guarded, in ages long past. The very high civilization which is now known to have existed thousands of years before Christ in the earliest Egyptian dynasties makes it almost inconceivable to imagine that subjects of such transcendent interest to mankind were not then part of the learning of the few, part of "the wisdom of Egypt." The seizure of this knowledge by the priestly caste and its restriction to themselves, with penalties to all intruders, was the natural consequence of the lower civilization that followed. Thus psychical phenomena became veiled in mystery, and ultimately degraded to a mischievous superstition. Mystic rites were added to impress the multitude; eventually, divination, enchantment, augury, and necromancy became methods of wielding a mysterious power held by the few. But such practices "wearied the people's intellect, destroyed their enterprise and distorted their conscience."[8] The industry and politics of the people became paralyzed by giving heed to an oracle, or to gibbering spirits, rather than to reason and strenuous endeavor. The great Hebrew prophets, the statesmen of their day, saw this clearly and had the courage to denounce such practices in unmistakable terms; warning the people that by using these things as an infallible guide, or as a religion, they were being misled, and reason was being dethroned from her seat. And so the burden of their speech was, "Thy spells and enchantments with which thou hast wearied thyself have led thee astray."[9] Hence these practices were prohibited, as a careful study of the whole subject shows, because they enervated the nation, and tended to obscure the Divine idea; to weaken the supreme faith in, and reverent worship of, the one omnipotent Being the Hebrew nation was set apart to proclaim. With no assured knowledge of the great world-order we now possess, these elusive occult phenomena confused both the intellectual and moral sense, and so they were wisely thrust aside. But the danger at the present day is very different. Instead of a universe peopled with unseen personalities, the science of today has gone to the other extreme, and as Mr. Myers once eloquently said, we are now taught to believe "the Universe to be a soulless interaction of atoms, and life a paltry misery closed in the grave." Were the Hebrew prophets now amongst us, surely their voice would not be raised in condemnation of the attempts we are making to show that the order of Nature contains an even vaster procession of phenomena than are

8 Prof. G. A. Smith in his brilliant and scholarly work on Isaiah, vol. 2., p. 199.

9 Cf. latter half of Isaiah, 47 ch.

now embraced within the limits of recognized science, and that behind the appearances with which science deals there are more enduring and transcendent realities.

I HAVE ventured upon this digression in the hope that I may remove the misgivings with which a part of our work is regarded by some leaders of religious opinion, who from time to time have been in communication with me. Perhaps I may also add that the aversion which some feel towards any enquiry into spiritualistic phenomena arises, I think, from a misapprehension. With what is spiritual, with religion, these phenomena have nothing in common. They may afford us a rational belief in the existence of life without a visible body, of thought without material protoplasm, and so become the handmaid of faith. But they belong to a wholly different order from that of religious faith. Our concern is solely with the evidence for certain *phenomena*; and as Professor Karl Pearson has said, "Wherever there is the slightest possibility for the mind of men to *know*, there is a legitimate problem for science." Hence *all* appearances, whether of microbes or of men, are legitimate subjects of investigation. Because they happen to be fitful, or phenomena occurring in an unseen environment, does not render the investigation improper or unscientific, though it makes it considerably more difficult.

Now the investigations we have published undeniably establish the fact that human personality embraces a far larger scope than science has hitherto recognized. That it partakes of a *two-fold life*, on one side a self-consciousness which is awakened by, and related to time and space, to sense and outward things; on the other side a deeper, slumbering, but potential consciousness, the record of every unheeded past impression, possessing higher receptive and perceptive powers than our normal self-consciousness, a self that, I believe, links our individual life to the ocean of life, and to the Source of all life. It is a remarkable fact that long ago the philosopher Kant instinctively stated the same truth. He says: "It is possible that the human soul even in this life stands in indissoluble community with all immaterial natures of the spirit world, it mutually acts upon them and receives from them impressions, of which, however, as man, it is unconscious as long as all goes well."[10] This, of

10 " Es wird kiinftig, ich weiss nicht wo oder wann, nocb bewiesen werden, dass die menschliche Seele auch iu diesen Leben in einer unaufloslich verknlipften Gemeinschaft niit alien immateriellen Naturen der Geisterwelt stehe, dass sie wechselweise in diese wirke und von ihnen Eindriicke empfange, deren sie sich als Mensch nicht bewusst ist, so lange alles wohl steht." (Kant's *Sammtliche Werke*, Hartenstein's Edition, 1867, vol. ii., p. 341.)

course, was Swedenborg's view. He frequently tells us, "Man is so constituted that he is at the *same time* in the spiritual world and in the natural world." Plotinus, who lived in the third century, held a similar belief; this was in fact the view of the Neo-Platonists and of the later mystics generally.[11] In connection with this subject, may I commend to you the perusal of Dr. Du Prel's *Philosophy of Mysticism*, which has been translated with loving labor by one of the earliest and best friends of our Society, Mr. C. C. Massey: perhaps the most valuable part of the work being the suggestive introduction which Mr. Massey has himself added.[12]

There is one interesting point in connection with spiritualistic phenomena that is worth a little attention. As we are all aware, the production of these phenomena appears to be inseparably connected with some special person whom we call "mediumistic." This fact affords perennial amusement to the man in the street. But from a purely scientific standpoint, there is nothing remarkable in this. Recent discoveries have revealed the fact that a comparatively few substances possess what is called radioactive power. Unlike ordinary forms of matter, these radioactive bodies possess an inherent and peculiar structure of their own. There is therefore nothing absurd in supposing that there may be a comparatively few persons who have a peculiar and remarkable mental structure differing from the rest of mankind. Moreover, the pathologist or alienist does not refuse to investigate epilepsy or monomania because they are restricted to a limited number of human beings.

Furthermore, physical science gives us abundant analogies of the necessity of some *intermediary* between the unseen and the seen. Waves in the luminiferous ether require a material medium to absorb them before they can be perceived by our senses. The intermediary may be a photographic plate, a fluorescent screen, the retina, a black surface, or an electric resonator, according to the length of those waves. But some medium formed of ponderable matter is absolutely necessary to render the actinic, luminous, thermal, or electrical effects of these waves perceptible to our senses. And the more or less perfect rendering of the invisible waves depends on the more or less perfect synchronism between the unseen motions of the ether and the response of

11 Vaughan's "Hours with the Mystics," vol. i., contains an excellent summary of the views of the Neo-Platoiiists. Philo Judaeus writing from Alexandria a few years B.C. says, "This alliance with an upper world, of which we are conscious, would be impossible, were not the soul of man an indivisible portion of the divine and blessed spirit." See also Thomas Taylor's translation of some of the works of Plotinus.

12 Here perhaps I may add one line expressive of my own indebtedness to and affectionate regard for my dear friend C. C. Massey, whose knowledge of all that relates to the higher problems before our Society is more profound than that of any one I know.

the material medium that absorbs and manifests them.

Thus we find certain definite physical media are necessary to enable operations to become perceptible which otherwise remain imperceptible. Through these media, energy traversing the unseen is thereby arrested, and, passing through ponderable matter, is able to affect our senses and arouse consciousness.

Now, the nexus between the seen and the unseen may be physical or psychical, but it is always a specialized substance or living organism. In some cases, the receiver is a body in a state of unstable equilibrium, a sensitive material—like one of Sir 's receivers for wireless telegraphy—and in that case, its behavior and idiosyncrasies need to be studied beforehand. It is doubtless a peculiar psychical state, of the nature of which we know nothing, that enables certain persons whom we call mediums to act as receivers, or resonators, through which an unseen intelligence can manifest itself to us. And this receptive state is probably a sensitive condition easily affected by its mental environment.

We should not go to a photographer who took no trouble to protect his plates from careless exposure before putting them in the camera. And I do not know why we should expect anything but a confused result from a so-called medium (or automatist, as Myers suggested they should be called) if the mental state of those present reacts unfavorably upon the sensitive. Infinite patience and laborious care in observation we must have (as in all difficult investigation), but what good results from any scientific research could we expect if we started with the presumption that there was nothing to investigate but imposture?

In connection with this subject of mediumship, it seems to me very probable that a medium, an intermediary of some sort, is not only required on our side in the seen, but is also required *on the other side* in the unseen. In all communication of thought from one person to another, a double translation is necessary. Thought, in some inscrutable way, acts upon the medium of our brain and becomes expressed in written or spoken words. These words, after passing through space, have again to be translated back to thought through the medium of another brain. That is to say, there is a descent from thought to gross matter on one side, a transmission through space, and an ascent from gross matter to thought on the other side. Now, the so-called medium, or automatist, acts as *our* brain, translating for us the impressions made upon it and which it receives across space from the unseen. But there must be a corresponding descent of thought on the other side to such a telepathic form that it can act upon the material particles of the brain of our medium. It may be even more difficult to find a spirit medium there than here. No doubt wisely so, for the invasion of our consciousness here might otherwise be so frequent and troublesome as to paralyze the conduct of our life. It is possible, therefore, that much of the difficulty and confusion of the manifestations which are recorded in our *Proceedings*, and in the very valuable contribution which Mr. Piddington has just given us of sittings with Mrs. Thompson, are due to inevitable difficulties in translation on *both* sides.[13]

Furthermore, if my view be correct, that the self-conscious part of our personality plays but a subordinate part in any telepathic transmission, whether from incarnate or discarnate minds, we shall realize how enormously complex the problem becomes. So that the real person whom we knew on earth may find the difficulty of self-manifestation too great to overcome, and only a fitful fragment of their thoughts can thus reach us.

There is, however, another view of the matter which to me seems very probable. The transition from this life to the next may in some respects resemble our ordinary awakening from sleep. The discarnate soul not improbably regards the circumstances of his past life "in this dream-world of ours," as we now regard a dream upon awakening. If, even immediately upon awakening, we try to recall all the incidents of a more or less vivid dream, we find how difficult it is to do so, how fragmentary the whole appears; and yet in some way we are conscious the dream was a far more coherent and real thing than we can express in our waking moments. Is it not a frequent and provoking experience that whilst some trivial features recur to us, the dream as a whole is elusive, and as time passes on even the most vivid dream is gone beyond recall? May it not be that something analogous to this awaits us when we find ourselves amid the transcendent realities of the unseen universe? The deep impress of the present life will doubtless be left on our personality, but its details may be difficult to bring into consciousness, and we may find them fading from us as we wake to the dawn of the eternal day.

Whatever view we take, the records of these manifestations in our *Proceedings* give us the impression of a truncated personality, "the dwindling

13 Miss Jane Barlow, who has made a close study of these communications, writes to me on this point: "The almost unimaginable difficulty in communicating may account for many of the failures, mistakes, and absurdities we notice. I think we are apt to lay too much stress on the want of memory. Apart from purely evidential considerations, there seems a tendency to regard it as a larger and more essential element of Personality than it really is. In my own case, for instance, any trivial cause—a headache, a cold, or a little flurry—scatters my memory for proper names. I can easily imagine myself forgetting my own name without suffering from any serious confusion of intellect in other respects, or the least decay of personality."

remnant of a life," rather than of a fuller, larger life. Hence, whilst in my opinion psychical research *does show us* that intelligence can exist in the unseen, and personality can survive the shock of death, we must not confuse mere, and perhaps temporary, survival after death with that higher and more expanded life which we desire and mean by immortality, and the attainment of which, whatever may be our creed, is only to be won through the "process of the Cross." For it is by self-surrender, the surrender, that is, of all that fetters "What we feel within ourselves is highest," that we enter the pathway of self-realization. Or as Tennyson expresses it;—

> "Thro' loss of Self
> The gain of such large life as match'd with ours
> Were Sun to spark—unshadowable in words,
> Them.selves but shadows of a shadow-world."[14]

14 So also Goethe:

> "Und so lang du das nicht hast
> Dieses: 'stirb und werde' !
> Bist du nur ein trtiber gast
> Auf der dunken Erde."

which a friend has rendered as follows:

> "Whoso heeds not this behest
> 'Die to win new birth.'
> Lives but as a hapless guest
> On a darkening earth."

Some Reminiscences Of Fifty Years' Psychical Research.[1]

By Sir William Barrett, F.R.S.

> "Science is bound by the everlasting law of honor to face fearlessly every problem that can fairly be presented to it." —Lord Kelvin.

Just fifty years ago this month I first began a serious and systematic investigation of psychical phenomena, and was so impressed with the supreme importance of the subject that so far as my time and strength permitted I have continued the investigation to the present time with unabated interest. Of course, other and far abler men both preceded me and worked contemporaneously with me; to some of these I will refer directly.

Early Psychical Researches.

By psychical research I mean the critical investigation, and, where possible, the verification of alleged supernormal phenomena, or of hitherto unrecognized human faculties.

So far as regards narratives of spiritistic phenomena, these of course go back to remote ages, and records are to be found in many different countries. Years ago Andrew Lang had several conversations with me on the value of the S.P.R. devoting some time to historical research on this subject, for he attached great importance to the fact of the wide diffusion, both in space and time, of phenomena similar to those we are now engaged in investigating.[2]

Biblical references show that Spiritualism was rampant in the early history

1 Read at a Private Meeting of the Society on June 17, 1924.

2 The laborious and admirable work by William Hewitt, entitled *History of the Supernatural*, is well known and should be consulted by all who are interested in the history of psychical research. It made a great impression on me when I read it some fifty years ago. The term *supernatural* should, however, have been *supernormal*, as all phenomena however novel and inexplicable they may appear to be—are really natural; only God is above and beyond Nature.

of the Jews; King Saul himself being a notable spiritualist, consulting his medium at Endor. As I have said elsewhere, before science had established a universal reign of law or of the great world order, the pursuit of these spiritistic phenomena was justly condemned by the ancient prophets as likely to lead to! intellectual and moral confusion. It seems probable, how- ever, that many of the prophetic writings were done automatically, as in the book of *Chronicles* David says that the instructions he gave regarding the building of the Temple were not his own ideas, for "the Lord made me understand in writing *by his hand upon me.*"[3]

One of the first psychical researchers of whom I can find any report was the learned and famous German Jesuit, Fr. A. Kircher. In his Latin folios, published in 1640, he discusses the cause of the motion of the' pendule explorateur' (a little ball or ring suspended by a string held by one hand) and of the forked divining rod, both of which at that time were the subject of acute controversy. Kircher showed that if the' pendule' or the rod were held, not by the hand, but by a rigid support, no motion occurs under any circumstances. He was thus led to discover the principle of unconscious muscular action, a discovery claimed two centuries later by the distinguished French chemist, Chevreul, and by Dr. W. B. Carpenter subsequently.

Some twenty years after Kircher's work appeared, one of the founders of the Royal Society, the Hon. Robert Boyle, 'the son of the Earl of Cork and the Father of s Chemistry,' in his *Philosophical Works* discusses the I question of the divining rod, and urges further experiments to test its value in the discovery of mineral veins, for the evidence he collected was conflicting. He remarks that eye-witnesses, who were far from credulous, told him of the great value of the rod, and one gentleman in whose hand it moved when he passed over a vein of ore, affirmed "that the motion of his hand did not at all contribute to the inclination of the rod, but that sometimes when lie held it very fast it would bend so strongly as to break in his hand."[4]

In one of the first volumes of the *Philosophical Transactions of the Royal Society* Boyle asks the question whether anyone could inform him "whether diggers do meet with the subterraneous demons which are said to inhabit the lead mines of Somersetshire, and in what shape and manner they appear?" In a subsequent number the famous Dr. Glanvil, also a Fellow of the Royal Society, replied that he lived near the lead mines in the Mendips, and that whilst the miners heard the knockings of the demons, and by following the directions of these knockings were led to plenty of ore, nevertheless he had

3 See 1 *Chron.* xxviii. 19.

4 Boyle's *Philosophical Works* (1738), vol. i., p. 172 and 173.

not found anyone who had actually *seen* the demons or gnomes themselves, and therefore he could not describe their appearance or habits.[5] A curious fact is that these knockings are heard and the gnomes believed in by lead miners in various parts of England and Ireland. The simple scientific explanation of these sounds I think I have discovered, but it would take me too far to go into the matter here. The point I want to bring out is that Boyle and Glanvil and many of the early Fellows of the Royal Society were true psychical researchers; as Prof, de Morgan has said, "they set themselves to work to prove all things, that they might hold fast to that which was good; they bent themselves to the question whether sprats were young herrings and whether a spider could crawl through the powder of a unicorn's horn." They enquired whether there was any value in magnetical cures and any good in Kenelm Digby's sympathetic powder. Even a century later the great Sir Isaac Newton describes in a series of letters—which I have seen, as they have been preserved by my friend, Mx. Blayny Balfour—how he spent much time and money in besting the value of certain alchemical powders which were said to turn lead into gold.

The early spiritualists really took the method of these pioneers of physical science. Though they might have been frequently too credulous, yet what they aimed at was to collect facts however improbable they appeared to be, and as De Morgan says,"The spiritualists beyond a doubt are in the track that has led to all advancement in physical science; they have the spirit and the method of the grand time when paths had to be cut through the uncleared forests in which it is now the daily routine to walk. Their spirit was that of universal examination, wholly unchecked by fear of being discovered in the investigation of nonsense."[6]

One of the earliest Fellows of the Royal Society, to whom I have already referred, was Dr. Joseph Glanvil he was chaplain to Charles II. and a prebendary of Worcester. The historian Lecky describes him as " a man of incomparable ability." Speaking of one of Glanvil's works, Lecky remarks,"it would be difficult to find a work displaying less of credulity and superstition than this treatise." Glanvil's *Saducismus Triumpliatus,* pubhshed at the end of the seventeenth century, discusses the evidence; concerning witches and apparitions, and gives the fullest report of those remarkable poltergeist phenomena known as the' Demon of Tedworth' or the "Disturbances in Mr. Mompesson's house in Wiltshire." Glanvil truly remarks that" matters of fact

5 *Phil. Trans.,* No. 19, Nov. 1666 ; and GlanvU's reply is in *Phil. Trans.,* vol. iii., 1668.

6 Preface to *From Matter to Spirit,* pp. 18-20.

well proved ought not to be denied because we cannot conceive how they could be performed. Deceit and fallacy will only warrant a greater care and caution in examining." Glanvil had the warm support not only of Robert Boyle, but also of the famous Henry More.

In his record of the hauntings of the Epworth parsonage in 1716, and in his endeavor to get first-hand evidence of other supernormal phenomena, John Wesley—as Mr. J.—Arthur Hill has said "would have made an excellent member of the S.P.R.," and it is certainly surprising—as Mr. Hill adds— "to find in an earnestly religious man of that day, such as John Wesley, so much critical interest in our subject."

Coming to more recent times, the memoir of the eleventh Duke of Somerset, the great grandfather of our friend and fellow-worker, Miss Ramsden, shows that the Duke was really a keen psychical researcher, for he critically examined several psychic cases, among others the well-known dream of a Cornish farmer, Williams, who on May 11, 1812, woke his wife and told her that he had dreamt he had seen a man shot in the lobby of the House of Commons, and described his appearance, etc. The dream was repeated, and told to several people; subsequently it was found that the details of the dream exactly corresponded with the facts connected with the assassination of Mr. Percival, the Chancellor of the Exchequer, on the eleventh May, the very evening Williams had his dream.[7] Percival's descendant—my venerable friend. Miss Percival" of Chobham—has also given me a written contemporary record of this remarkable dream.

Among the Fellows of the Royal Society who warmly advocated the investigation of psychical phenomena, was Dr. Mayo, F.R.S., Professor of Physiology in King's College, London, whose writings on the subject, published about the middle of the last century, are well worth perusal, containing as they do many new and interesting facts, which at the present day seem little known.

This period was also notable for the great interest excited by mesmeric phenomena. The remarkable report of the committee appointed by the medical section of the French Academy of Sciences on this subject, together with the amazing phenomena to which they testify, excited widespread interest. Eminent English physicians and surgeons, such as Dr. Elliotson of St. Thomas' Hospital and Dr. Esdaile, presidency surgeon in Calcutta, made numerous contributions to our knowledge on this subject, especially as regards the therapeutic and analgesic power of mesmerism. Esdaile, as is

7 See p. 335 in Lady Gwendolen Ramsden's *Correspondence of Two Brothers from 1809-1819* (Longmans & Co.).

well known, conducted a very large number of *major* surgical operations, absolutely painlessly, under the mesmeric trance and if the use of chloroform as an anaesthetic had not been discovered about this time, the value of mesmeric trance in surgical operations would have been universally acknowledged. But in spite of the eminence of Dr. Elliotson and others, mesmerism was nevertheless denounced by the profession as a whole, and the *Lancet* called it "an odious fraud." It was not until Braid of Manchester employed the word *hypnotism* instead of mesmerism—and thus dissociated the subject from Mesmer, who was more or less of a quack—that the medical profession began to treat the subject with less contempt. Braid also gave a reasonable explanation of the phenomena, but subsequently this proved to be inadequate to account for all the facts. The valuable experimental work of the continental hypnotists, and of our own members, Edmund Gurney, Dr. Lloyd Tuckey and Dr. Milne Bramwell, have now placed hypnotic treatment among the recognized therapeutic agencies of the medical profession.

Foundation of the Society for Psychical Research.

Some twenty-five years before our Society was founded a few of the younger Fellows of Trinity College, Cam- bridge, chief of whom was Mr. B. F. Westcott—afterwards the famous Bishop of Durham—started a' Ghost Society' very much on the lines of our S.P.R. In our *Journal* for April last year I gave the object and the brief history of this Society, which included among its members several Cambridge graduates who subsequently became eminent, such as Archbishop Benson, Prof. Henry Sidgwick, and others. I will hand over for preservation among the archives of our Society the original document of the 'Cambridge Ghost Club' (as it was called), given to me by my friend the Archdeacon of Dublin.

As regards the foundation of our own Society, I have corrected elsewhere the misunderstanding which Prof. Richet and others have fallen into;[8] and also I wrote, by request, a fuller account of the early history of our Society. [9]Though it is true I happened to be the chief instrument in the foundation of our Society in 1882—and of kindred societies in Canada and the United States in 1884—yet the high position and respect the S.P.R. has won is chiefly due to Sidgwick, Myers, and Gurney, the three great pillars upon which the edifice of our Society was originally built. These eminent men were unlike in many ways: Sidgwick by his adhesion to the S.P.R. greatly impressed the academic world, for his influence, wisdom, and caution were

8 See *Journal of the S.P.R.,* vol. xxi., October 1923, p. 139.

9 See *Light*, June 21, 1924.

widely recognized; Myers by his enthusiasm, brilliant talents, and profound intuition, was the corner stone of the S.P.R. till his death in 1901: Gurney with his industry and immense range of knowledge was essential to the early progress and stability of our Society.

Crookes was really the first scientific man to devote his experimental skill, from 1870 to 1874, to the critical investigation of the physical phenomena of spiritualism. It is pitiful to think of the scientific ostracism to which he was subjected and over which his genius eventually triumphed. I am not, however, surprised at the Royal Society refusing to publish his spiritistic investigations, for, a few years later, my modest paper read before the British Association in 1876 was refused publication by scientific societies, of which I was member, on the very natural ground that science dealt with the evidence furnished by our recognized senses, whereas my paper dealt either with . phenomena which transcended the usual channels of sense or with phenomena that transcended the material world.

Early Steps in Psychical Research.

Many friends have asked me how I first became interested in psychical research. Perhaps I may be forgiven for relating an old story. Between the years 1862 and 1867 I was assistant to Professor Tyndall at the Royal Institution. The atmosphere surrounding my early years there was entirely opposed to any belief in psychical phenomena. Faraday, to whom electrical engineers owe the source of all their vast undertakings—yet who lived and died a poor man—Faraday I saw almost daily, before he left the Royal Institution and went to Live at the Hampton Court Cottage given to him by the Prince Consort. I can never forget the debt I owe that famous man for his generous kindness and guidance to a young ignoramus Like myself. Faraday had published about 1855 his famous experiment on table-turning, showing how unconscious muscular effort accounted for what *he* saw. A little later he publicly declined to sit with the medium Home, saying he had lost too much time over such matters already. Tyndall also had denounced spiritualism as an imposture. Both Huxley and Herbert Spencer were frequent visitors to the Royal Institution laboratory, and both of these eminent men treated all psychical phenomena with contemptuous indifference. Among other frequent visitors was an Irishman, Mr. John Wilson, who invited me to spend my vacations at his place in County Westmeath. For a couple of years I did so, and found to my astonishment that Mr. Wilson was a firm believer in—and experienced investigator of—mesmerism, as it was then called. He showed me some most extraordinary experiments upon a sensitive subject from his estate. I was naturally in- credulous and asked to be allowed to repeat the experiments myself, selecting another subject. We found a young uneducated Irish girl, who proved to be extremely sensitive. In the mesmeric trance—in spite of every pre- caution that I took to prevent deception—whatever sensations I felt, whether of touch, taste or smell, were transferred to the subject, and, moreover, ideas and words which I thought of were reproduced more or less accurately by the hypnotized subject. A brief account of some of these experiments may be found in the first volume of our *Proceedings*, p. 240.

When I returned to London I repeated these experiments with a couple of boys whom I found susceptible to hypnosis. These experiments revealed to me the extra- ordinary power of either verbal or silent suggestion upon the hypnotized subject. For instance, placing a pair of shoes upon the table, I told one of the lads that I was going to float round the room and pointed to the position I was supposed to have reached near the ceiling. Then, clapping my hands, I suggested that I was safely back in my shoes on the table. The boy stared at me with incredulity and related afterwards, both to me and to others, that he had really seen me floating round the room. A similar result was obtained by me with another subject when I was on a visit to America in 1884. Hence I was led to believe that spiritualistic phenomena, when not fraudulent, were really due to the *hallucination* of the observer; that, in fact, the phenomena, such as Home floating out of the room or putting his hands into the fire, were really *subjective* and not objective. I had been in correspondence with Mr. Crookes on scientific matters so far back as the year 1864, when I published in the *Philosophical Magazine* my first scientific research. In 1870 Crookes began his experiments with the famous medium D. D. Home; soon after this he wrote to me (in a letter which I have kept) as follows::

"*May* 14, 1871.

Dear Mk. Barrett, I must have some conversation with you respecting these obscure phenomena. If you can help me to form anything like a physical theory I should be delighted. At present all I am quite certain about is that they are *objectively* true. I have had all my wits about me when at a seance, and the only person who appeared to be in a state of semi-consciousness is the medium himself. The other evening I saw Home handling red-hot coals as if they had been oranges. Will you favor me with a visit one evening when you are disengaged. Very truly yours,

William Crookes."

I had several interviews with Crookes, but unfortunately lad no opportunity of any sitting with Home, and a year Or two later I was appointed to the Chair of Physics at the Royal College of Science, Dublin. In 1874 I made my first acquaintance with the physical phenomena of spiritualism, and was able to put to the test my preconceived theory of hallucination, which was gradually dispelled, and I became convinced of the *objective* reality of the phenomena. It so happened that one of the London weekly reviews had sent me about this time a number of books on spiritualism to review, and in a lengthy review I suggested the hallucination hypothesis, but in a postscript added that this view was open to serious doubt.

The record of my experiments in Ireland on the physical phenomena of spiritualism was read before the S.P.R., and will be found in our *Proceedings*, vol. iv., p. 25, etc. I was singularly fortunate in these early experiments on physical phenomena, as the mediums were personal friends, and the experiments took place in *full light* either in my own house or in that of my friends. After this lapse of time I think I may mention without indiscretion that the name of the young medium I called Florrie was Miss Clark; her father, a leading London solicitor, had taken a furnished house near my residence in Kingstown. The house belonged to Mr. James Wilson (brother of my Westmeath friend), who asked me to call on Mr. Clark. Mr. Wilson was the father of the late Commander-in-Chief, Sir Henry Wilson, an old student of mine. My prolonged investigation of Florrie Clark, in *full daylight*, and other experiments with the family of the Landers (one of the leading pho-to- graphers in Dublin), so impressed me with the supreme importance of the whole subject that I determined to let no opportunity pass of pursuing these investigations. Shortly afterwards I prepared a paper on the various psychical phenomena I had witnessed ; this was accepted by and read before the British Association in 1876.[10]

Verbatim reports of my paper were given both in the local and the spiritualist journals of Sept. 1876.[11] It will] be seen from the reports of this paper that, while some notable scientific men such as Dr. W. B. Carpenter vigorously opposed my paper, others still more notable spoke ir my support;

10 The greater part of this paper (which the British Association refused to publish) will be found in *Proceedings of the S.P.R.*, vol. i., p. 238 *et seq.*

11 These are preserved in a scrap-book, which I still have.

such, for example, as Mr. (afterwards Su William) Crookes and the late Lord Rayleigh (both of whom subsequently became Presidents of the Royal Society), also the president of the Section, Mr. A. R. Wallace, and the distinguished anthropologist Col. Lane Fox, who afterwards became General Pitt-Rivers. I may add that a warm letter in support of my paper was received by me from the famous astronomer who subsequently became Sir William Huggins, also a President of the Royal Society.[12] The main object of my paper was a plea for a scientific committee to examine the validity of the evidence that I had adduced on behalf of what appeared to be thought-transference and other supernormal phenomena, especially those relating to spiritualism. Unfortunately, in the violent dispute which followed between Carpenter and Wallace, my resolution was lost sight of.

For weeks a great controversy ensued in the London *Times*, which, like all the other newspapers (with the exception of the *Spectator*), poured ridicule upon my daring to bring such a contemptible subject before the British Association. Among other of my vigorous opponents in *The Times* were Professors Lankester and Donkin (now Sir Ray and Sir Bryan); both of these distinguished men are still living and have not abated their hostility to the subject. Here it is interesting to note that in 1876 I ventured to state in *The Time*s that before we could hope to arrive at any definite conclusion as to the origin of spiritualistic phenomena, we must first ascertain whether such a thing as the transfusion of thought between sitter and medium really existed and its extent; and, secondly, whether such a thing as clairvoyance or a transcendental perceptive power had any foundation in fact? Both these questions have now been answered in the affirmative, ind this renders the above caution the more necessary.

'Book tests' show that Richet's theory of cryptesthesia, which seems to be another name for clairvoyance, needs serious consideration. The critical examination of Stainton Moses' script by our member Mr. Trethewy has shown that many apparently spirit communications are transcripts of some matter which had been written or printed, it may be long ago, and, as a rule, only to be found in places inaccessible to the medium. *Nothing* seems able to elude the prying eyes of the transcendental self, whether it be incarnate or discarnate. In many sittings of to-day the same thing holds good.[13] For example, that excellent but illiterate medium, Sloan, of Glasgow, when m

12 It is interesting to note that of the few scientific men who then had the courage to support my B.A. paper, no less than four received the Order of Merit.

13 The most amazing illustration of telesthesia, or travelling clairvoyance, I have ever known I hope to read before the S.P.R. shortly.

a trance state will often give details concerning a sitter whom he has never seen before and whose name even he does not know, yet in subsequent investigation many of the facts stated by the medium will be found printed, either in *Who's Who* or other books to which the medium appears to have had no possibility of access.

Hence, with these and other facts known to us, we must realize that, however trustworthy may be the *evidence* we obtain of supernormal phenomena, the *interpretation* of that evidence may in time alter—as our experience grows wider, and our knowledge of human psychology more extensive and profound. Albeit I am personally convinced that the evidence we have published decidedly demonstrates (1) the existence of a spiritual world, (2) survival after death, and (3) of occasional communications from those who have passed over.

It will be obvious that in the present paper I cannot give even an outline of the evidence that has led me to the foregoing conclusions. The so-called physical phenomena afford striking evidence of amazing super-normal power, but no proof of the survival of human personality can be derived from them. In fact, many people believe that they are simply due to the 'psychic force' of the medium and sitters; however, this is an opinion I do not share. Even the so-called 'ectoplasm,' which is seen issuing in certain cases from the body of the medium and sometimes taking human shape, may be attributed to some mysterious, unknown, and unconscious power possessed by the medium. There can be little doubt that the source of the ectoplasm is m some way derived from the human organism; but I believe an unseen and intelligent supernormal agency *external* to the medium causes this ectoplasm to take definite forms and apparent vitality. I myself have not had the opportunity of ever witnessing these ectoplastic phenomena, and therefore I defer to the opinion of such experienced and critical observers as the late Dr. Geley, Prof. Richet, and others.

Physical Phenomena

As regards the ordinary physical phenomena I have had a long series of experiments with various mediums, going back to the time of Dr. Slade in 1876. Slade always sat with me in broad daylight, and though I have little doubt that he not unfrequently resorted to trickery, yet there was also no doubt he had genuine and remarkable mediumistic powers ; his so-called exposure by Prof. Ray Lankester was quite inconclusive. Those who are interested in the matter will find in Stainton Moses's *Psychography*, pp. 104-

110, a record of some of my experiments with Slade, in part quoted from the London *Times* of September 1876, and also a careful report written by my friend Mr. Conrad Cook, who accompanied me to a sitting with Slade in August 1876.

The only case of 'materialization' witnessed by me, which seemed to be inexplicable by fraud, occurred with the medium. Husk, many years ago. It may be worth while describing this experiment as it has never been published.

Mr. Wm. de Morgan had kindly lent Myers and myself ills studio in Cheyne Row, an almost bare room, furnished with a small deal table about 3 feet by 5 feet, and a few chairs. After dinner Myers brought Husk to Cheyne Row in a hansom cab, and we immediately sat round the table. There were six present including the medium. Wilham de Morgan and his sister (being skeptics), were placed in control of the medium, whose feet were tied to the legs of the table, and his hands were grasped by the sitter on each side. Mrs. de Morgan (their mother), at facing Myers, and I sat at the other end of the table and had control of the light. After the wrists of all present had been loosely joined together by silk thread, I blew out the candle and phenomena very soon occurred. The medium went into a trance, lights, very like fireflies, were seen darting about over our heads, movement of some objects in the room was heard, and a deep guttural voice spoke to us calling himself 'John King.' In reply to our request he said he would try and show himself. A violent convulsion of the medium occurred, and suddenly right in front of me appeared a clothed human figure from the waist upwards : the lower part of the body might have been concealed by the table. The face was illuminated by a bluish light which seemed to issue from an object held in the hand of the materialized figure. The face was undoubtedly a living one, for I saw its eyes open and close and its lips move; I asked who it was and the guttural voice said "John King." It was a dark bearded and rather unpleasant face, quite unlike that of the medium. I exclaimed, " Do you all see the figure? I am going to light the candle," and immediately risked doing so. The figure vanished the moment the match was struck, and the medium was found in deep trance, lying back in his chair and groaning: when the medium had recovered he was sent home in a cab. On comparing notes, each sitter described the face according to the different aspects it presented from his or her position at the table. We found upon experimenting that it was impossible to reproduce the figure by leaning over the table, nor could the medium have put on a mask, as his hands were held the whole time and the tying of his legs and wrists were found intact. De Morgan asked Myers and myself to come the next morning and see if we could in any way imitate what we had seen. Though de Morgan remained somewhat skeptical, Myers and I both agreed that it was extremely difficult to explain the phenomena by trickery on the part of the medium, who, moreover, was found deeply entranced a few seconds later.

With regard to the so-called 'spirit photographs,' I have been extremely skeptical of their genuineness until quite lately. Recently, however, experiments conducted by my friend Mr. de Brath, in one of which he kindly allowed me to take part, appear to afford *indubitable* evidence of supernormal psychic photography. This conclusion confirms the opinion held by some expert and critical experimenters, who have discussed their results with me. Of course faked 'spirit photographs' abound and are easy to produce; whilst heartless rascals exist who prey upon the grief of a credulous sitter. Healthy skepticism has therefore been inevitable and wise. But we shall never arrive at any knowledge of the conditions requisite for these and other marvelous psychic phenomena, until hostile incredulity becomes no longer possible. Then, as Sir John Herschel says, "occurrences which, according to received theories *ought not* to happen, are the *facts* which serve as clues to new discoveries."[14]

These disputable subjects illustrate the importance of our society recognizing the fact that a difference of opinion—a *right* and a *left* wing—will necessarily have to exist among its different members. I mean that there are some who have been convinced at first hand, from their own experience, that the existence of certain psychical phenomena—especially those associated with spiritualism—admit of no doubt whatever, and are impatient with those who have not had this experience and are therefore more inclined to be cautious and even skeptical. The former class of our members wish to push forward and perhaps attach less importance to conclusive experimental evidence than they did at first : the latter class wish to go much more slowly and proceed step by step. This difference of opinion, though healthy, naturally leads to a divergence of interest in our subject, and from time to time threatens to break up the solidarity of our society.

Nor must we forget that psychical research, as stated in the original articles of our society embraces far more than spiritistic phenomena, and I hope that our research officers will ever bear in mind the varied objects of our society, which will be found printed in the first volume of our *Proceedings.*

Personally I am very anxious that earnest attention should be given to the so-called 'Reichenbach Phenomena,' wherein certain sensitives after long

14 *Discourse on Natural Philosophy*, section 5.

immersion in complete darkness perceive a luminosity emanating from the poles of a magnet and also from the human fingers. I have published in the *Philosophical Magazine*, and also in the early volumes of our *Proceedings*, the experiments which led me to the conviction that such phenomena do really occur under suitable conditions.

Another point, which I hope will sooner or later be the subject of further experiment, is the question of the old mesmeric hypothesis of 'effluence,' for which both Gurney and myself obtained what appeared to be satisfactory evidence forty years ago, and which Prof. Alrutz has in recent years confirmed.

The immediate work before us is to convince scientific authorities that various types of supernormal phenomenal do *really* exist, and are capable of experimental proof I do not think that the indifference of official science to our investigation, which has now replaced their former hostility, will be affected by an appeal to the *emotions,* such as the evidence obtained of survival after death. Science will approach the supernormal from an entirely different angle; it may possibly be pulled over the line of its present indifference by a growing recognition that telepathy does really exist. Unfortunately, the difficulty of finding suitable subjects both for telepathy and telethesia is a great drawback. For this reason I am led to the conviction that the first movement of thought of official science in our direction will occur from the impossibility of finding any *normal* explanation of the phenomena of *dowsing.* And this subject is of all others the simplest and easiest to investigate. It has not only a wide practical importance, but it raises no religious opposition, even from the most timid of narrow-minded people. Moreover, the number of efficient dowsers can easily be ascertained and experimented with. Having devoted more than twenty years to the critical investigation of this subject, I believe that it affords the easiest and most conclusive evidence that a supernormal perceptive power—akin to clairvoyance—exists in certain persons, of either sex, of all ages and all degrees of education.

This faculty is to be found not only in various races of men, but appears to exist, as a new and necessary sense, in many of the lower types of life. I am convinced that the mysterious migration of birds, even of very young ones, over vast tracts of land and sea, and also the *homing* instincts of many birds and animals, will be found to be akin to the dowsing faculty in man.[15]

15 See Chapter Two of my little book *Creative Thought,* published by Vatkins, Cecil Court, London, W.C.

Difference between Physical and Psychical Research.

There is, of course, a fundamental difference between physical research and psychical research. The former deals with matter and energy, and the condition of the mind of the observer, whether he be skeptical or not, is of no consequence. The latter deals with the phenomena of the subconscious, and the mental attitude both of the experimenter and his subject, is of prime importance. An interesting illustration of the difference between the mental states in the two cases occurred to me some time ago. I happened to be staying in Edinburgh with that famous physicist, Professor Tait, when the news of the discovery of the telephone came to us by cable. I asked Tait what he thought of it. He replied, "It is all humbug, for such a discovery is physically impossible." When I asked him low it was that well-known men had asserted that they lad heard speech transmitted a couple of miles through wire, Tait replied it was "probably a case of the conduction of sound by long straight wires." A little later, when the telephone was shown at the British Association by Sir W. Thomson (afterwards Lord Kelvin) and experiments with it were successfully made, Tait's obstinate incredulity did not interfere with the success of the experiment. A similar instance occurred in Paris when the Abbe Moigno, a well-known scientific writer, first showed Edison's phonograph to the Paris Academy of Sciences; the Abbe himself related to me what occurred. All the *savants* present declared, as Professor Tait did, that the reproduction of the human voice by an iron disc was physically impossible owing to the subtle wave forms produced by speech, though they admitted music might be so transmitted. The Abbe was even accused of having a ventriloquist concealed beneath the table. He left the chamber in disgust and told them to try the instrument themselves—which they did quite successfully in spite of their utter incredulity.

How different is the effect that is produced by mental environment in psychical phenomena is well known. Those experienced observers, Stainton Moses and C. C. Massey, have said, "the most unfavorable disposition to take to a medium is suspicion, and the most favorable is confidence." Skeptics may think that this is to deliver oneself over as a prey to the deceiver, and some men do certainly get taken in, but experience after a time leads them to discover their mistake. In the psychical world "faith," as Mr. Massey says, "is a condition of obtaining evidence, the key to the gate of the invisible world." By faith Massey means that a *sympathetic* state of mind establishes a rapport between the observer and the medium. The fact that we need no sympathy with our instruments, when testing a physical or chemical discovery, naturally leads the physicist or chemist to a state of scornful amusement, when told that his own attitude of mind is of importance in psychical research. But it appears to be a fact that even if at the back of our minds we entertain feelings of prejudice and hostility, we can hope for little success in psychic enquiry, however much we may disclaim the feeling of hostility. This, of course, does not mean an attitude of credulity or any relaxation of careful and critical observation.

All psychical researchers need to bear in mind that every sensitive or medium is a *suggestible* subject; if you go expecting fraud you may possibly create the very fraud you suspect. If you make preparations beforehand to lay a trap for the medium, it is probable that both medium and experimenter will fall into the trap.

There is another aspect of our enquiry, known to most of us—that is, that psychical phenomena largely depend upon involuntary and not voluntary effort, upon the sub-conscious and not the conscious self. Even in the simple phenomena of telepathy it is the subliminal self that is operative. Further, I believe that the common practice of experimenters *energetically willing* the idea to be thought of, is of no value, and may indeed be detrimental to success. In the early experiments which Myers, Gurney, and myself conducted with the Creery children at Buxton we found that the best results occurred when no strenuous efforts were made. In fact, when we made the experiments as amusing as possible, we had the greatest success, though every precaution was taken to prevent collusion or signaling.

Another fact which seems to me brought out very clearly in our experiments is that psychical phenomena, whether of telepathy, clairvoyance, or the higher phenomena of spiritualism, are manifestations of, or through, the *transcendental* self of the subject, and are therefore independent of the fundamental units of the physical world—matter, time, and space. It is true that, in the case of telepathy, the mental response of the percipient to the idea in the agent's mind, naturally suggests the physical analogue of the resonance of a silent tuning-fork to a sounding one which is in perfect unison with it. Indeed, I was inclined it first to think that telepathy was somewhat similar to this—that it was a *nervous induction* across space, analogous to the well-known facts of electric and magnetic induction. But whilst telepathy has been made more conceivable, and more credible to the public generally, by the discovery and use of wireless telephony, we must remember that the two phenomena are wholly different. One belongs to the *physical* order, the other to the *psychical* order. The laws regulating the transmission of energy across space apply to the one, but not to the other. Immense effort is necessary to transmit a wireless message across the Atlantic,, but apparently no effort at all is required to transmit a telepathic impact, of which we have instances, from New Zealand to London. On the contrary, a *passive* condition of both transmitter and receiver in telepathy seems essential, so far at least as their consciousness is concerned.

The word 'thought-transference' is apt to be mis- leading, as it seems to suggest a transmission of ideas between two persons across material space; but, as I said, space does not seem to enter into the question at all. Here it may be interesting to note that in the first publication of the discovery of this super-sensuous faculty, I called it not 'thought-transference,' but *the transfusion of thought.*[16] We are now coming back to this idea, for telepathy is probably the intermingling of our transcendental selves or souls. The common and grossly materialistic conception of the soul is that it is limited to the confines and contour of the body. This is surely an erroneous conception if, as we believe, the soul is an *immaterial* entity, not simply a function of the brain. For all we know to the contrary, the human soul may spread through a vast orbit around the body, and may inter- mingle with other incarnate or discarnate souls. Tennyson speaks of a dream condition, "when the mortal limit of the self was loosed, and past into the Nameless, as the cloud melts into Heaven." Moreover the intimacy and immediacy of the union between the soul and God is the fundamental idea, not only of the New Testament, but of all great Christian thinkers.

Conclusion.

As evidence of the great value which some eminent men attached to our investigations at their very outset, I will only quote from a couple of letters which I received more than forty years ago. That distinguished scientific man of the last generation. Dr. Angus Smith, F.R.S., writing to me in 1876, on the theoretic importance of thought-transference, remarked that "the indications now obtained point to some mighty truth more decidedly than even the aberrations of Uranus to the newest of the great planets. If we could prove the action of mind at a distance by constant experiments it would be a discovery that would make all other discoveries seem trifles." This was also the view of that eminent biologist, Mr. G. J. Romanes, F.R.S., who, when writing to me on the same subject in 1881, remarked "if the alleged phenomena are

16 See *Proceedings,* S.P.R., vol. i., p. 48, where will be found an extract from a letter of mine to *The Times* dated Sept. 1876.

true I hold it to be unquestionable that they would be of more importance than any other in the science and philosophy of our time."'

Quite recently our former President, Professor W. M'Dougall, F.R.S., in his presidential address to the American S.P.R. speaks of psychical research as the most hopeful barrier against the oncoming tide of materialism, and he remarks that " a civilization which resigns itself holly to materialism lives upon and consumes its moral capital and is incapable of renewing it. . . . Unless psychical research can discover facts incompatible with materialism, materialism will continue to spread; *no* other power can stop it, both revealed religion and metaphysical philosophy are equally helpless before the advancing tide. As regards religion being helpless, I cannot, however, go as far as M'Dougall.

Richet's point of view, which is purely materialistic, appears at first to contradict M'Dougall's remarks. We know that Richet, with splendid courage and loyalty to truth, has avowed his belief in the most incredible psychical phenomena, some of which even we may perhaps hesitate I accept. But Richet's philosophy compels him to reject thee spiritualistic hypothesis and to explain everything by a modified psychic force theory; a theory which was once accepted by Crookes but subsequently rejected by him. Richet attributes all the subjective phenomena of psychical search to 'cryptesthesia,' and some of the objective to 'pragmatic cryptesthesia.'[17] But these polysyllables do not help us any more than the names given by some learned psychologists who tell us that all psychical phenomena are simply illustrations of the "exteriorised effects of unconscious complexes!" One is reminded by this formidable nomenclature of the numerous and recondite hypotheses by which Ptolemaic astronomers tried to make their observations square with the geocentric theory of the universe. To the plain man it seems simpler, less improbable and more in accordance with facts, for biologists to recognize—what astronomers long since have done—that the universe after all is *not* explicable from the restricted view-point either of the earth or of the brain. Nevertheless, Richet's views will doubtless form the half-way house of many *savants* who hold mechanistic theories of the universe. However, I venture to predict that neither they not Richet will remain many years in that convenient but anomalous resting-place.

Sooner or later psychical research will demonstrate to the educated world, not only the existence of a *soul in man*, but also the existence of a *soul in*

17 Richet uses this term instead of psychometry (soul measurement), which he rightly says is so detestable a word that he proposes to call xxx "pragmatic cryptesthesia, i.e. cryptesthesia by means of materialxxxxxxects."

Nature. Our biologists have hitherto been so largely wedded to materialistic views that they have overlooked the vast importance of the psychic factor in evolution. The recognition of such a purposive and a pervasive factor, running throughout the whole realm of nature, will be found necessary to invoke in order to explain many biological phenomena that now receive very inadequate solutions from current theories. Long ago Lord Kelvin said, "Overpoweringly strong proofs exist of intelligence and benevolent design in Nature."

At the present day, when the very foundations of religion appear to be shaken, and men are deserting the faith of their fathers, and the whole civilized world is becoming more and more materialistic in its views, it is evident that psychical research will ere long be regarded, by any thoughtful men, as the most valuable handmaid to religion. Scarcely a week passes without my receiving letters or visits from perplexed men, both among the clergy an laity, who have found their religious creeds crumbling beneath their feet, and want to know what help they might obtain from psychical research. Mr. Gladstone's opinion on this subject is well known and often quoted; and Frederic Myers, as we know, has eloquently expressed his views. In his *Human Personality* again and again he returns to this aspect of the subject, and in his last chapter remarks, "We do not seek to shape the clauses of the great Act of Faith, but merely to prove its *preamble*. . . To be able to say to the theologian or philosopher: Thus and thus we demonstrate that a spiritual world exists—a world of independent and abiding realities, not a mere epi-phenomenon or transitory effect of the material world—but a world of *things,* concrete and living, not a mere system of abstract ideas' . . ." and he adds, "This would indeed, in my view, be the weightiest service that *any* research could render to the deep disquiet of our time—to the world-old, and world-wide, desire"[18] of mankind.

18 *Human Personality*, vol. ii., p. 297.

INDEX

Death-Bed Visions was first published in 1926. Other texts in this collection first appeared in the years noted. The text of this new edition was produced from copies of the first printings and has been reset by the Principle Press using the classic typeface *Garamond*. Founded in 2019, the Principle Press is dedicated to reviving and preserving important and often overlooked printed works with care and integrity.

Made in United States
North Haven, CT
04 November 2026

21767233R00125